Control Based on PID Framework - The Mutual Promotion of Control and Identification for Complex Systems

Edited by Wei Wang

Published in London, United Kingdom

IntechOpen

Supporting open minds since 2005

Control Based on PID Framework - The Mutual Promotion of Control and Identification for Complex Systems
http://dx.doi.org/10.5772/intechopen.87696
Edited by Wei Wang

Contributors
Ibrahim Mohammed, José Pinheiro De Moura, João Viana Fonseca Neto, Shin-Horng Chong, Allan Roong Soon Chan, Norshaslinda Hasim, Damir Vrancic, Mikulas Huba, Piotr Wladyslaw Ostalczyk, Piotr Duch, Jinping Feng, Wei Wang

First published in London, United Kingdom, 2021 by IntechOpen
IntechOpen is the global imprint of INTECHOPEN LIMITED, registered in England and Wales, registration number: 11086078, 5 Princes Gate Court, London, SW7 2QJ, United Kingdom
Printed in Croatia

British Library Cataloguing-in-Publication Data
A catalogue record for this book is available from the British Library

Additional hard and PDF copies can be obtained from orders@intechopen.com

Control Based on PID Framework - The Mutual Promotion of Control and Identification for Complex Systems
Edited by Wei Wang
p. cm.
Print ISBN 978-1-83968-366-4
Online ISBN 978-1-83968-367-1
eBook (PDF) ISBN 978-1-83968-368-8

For EU product safety concerns:
IN TECH d.o.o., Prolaz Marije Krucifikse Kozulić 3, 51000 Rijeka, Croatia, info@intechopen.com or visit our website at intechopen.com.

Meet the editor

Dr. Wei Wang received his Ph.D. degree in control theory and operations research from the Institute of Systems Science, CAS, China, in 1994. He is now a professor at the School of Mathematics, Renmin University of China. As the co-author of the paper "Nonlinear tracking-differentiators," he and his Ph.D. advisor Professor J.Q. Han won the 1979-2009 Best Paper Award for System Science. He proposed a series of innovative methods, such as "new separation principle of PID control," "double adjustment mechanism," and "disturbance stripping principle" for the control of uncertain networked systems and even complex systems. He proposed the estimation methods for the time-invariant and time-varying parameters of nonlinear systems and the adjustment method on the adaptability of the time series model based on uncertain system control theory. He also proposed the new strategy on the reconstruction or prediction for observations or signals with uncertain dynamics based on Taylor's formula and the theory of tracking-differentiators. Combining this strategy with the self-interaction characteristics of polymers, he proposed a personal growth model based on the dialectical relationship between internal and external factors, and the spatial-temporal model for the evolution of social systems based on the theory of fractal geometry and diffusion processes. He published more than 100 papers and 2 books.

Contents

Preface

Proportional–integral–derivative (PID) control, because of its simplicity in structure and its robustness with the uncertainties and disturbances, is such a popular control technology even today and has been widely used in many industrial processes. Furthermore, from a philosophical viewpoint, PID control is the embodiment of such a control idea that the control input can be determined based on the combination of the information at the present (P), the past (I), and the future (D), which is exactly the representation of the human being's wisdom, idea, and method to problem-solving.

Due to the popularity of PID control, a ton of content addressing this topic has already been made available by various publishers. At the same time, countless optimization opportunities have emerged with the development of science and technology, such as big data, artificial intelligence, machine learning, etc. There are also plenty of new fields that can leverage the power of control technologies. Some of the great examples include biological engineering, medical technologies, and financial technologies. The list is even extended to applications for social management and/or regulations, etc.

Therefore, PID control will continue to embrace new opportunities and face new challenges at the same time. This book highlights some of the new aspects of this subject and its related applications. Generally speaking, the better the cognition, the easier the realization of control. Yet, the relationship between cognition and control of complex systems can go both ways. We also know that the proper control method may be helpful for the improvement of the cognition of systems. That is to say, there is a mutual promotion between control and identification of complex systems that happens to reflect the law of the spiral rise in independent cognitive processes.

The main topics of this book include the improvement of PID controller, intelligent PID controller, optimization of PID controller, new forms of PID controller, development of PID controller, new application fields, etc. On the other hand, this book will also reveal that the proper control with a PID framework can also improve the cognition or identification for complex systems.

We intend to not only summarize the advance of science and technology on PID control but also extend the connotation and its extensions. We would like to bring more inspiration and benefits to the new fellows.

Based on the previous considerations, we present the book with the following chapters.

In Chapter 1, for some complex systems, the novelty method of the variable, fractional order PID controller is proposed. The main idea is to apply the split orders to discrete differentiation and summation functions, such that in the final time interval, the variable, fractional order PID controller must transform itself to the classical one that preserves the well-known stability conditions and zero steady-state error signal.

Chapter 2 focuses on the design and simulation of a hybrid LQR-PID controller used to stabilize the elevation, pitch, and travel axes of a helicopter system. An improvement in the performance of the hybrid LQR-PID controller is achieved by using a genetic algorithm (GA), which is adopted to obtain the best values of gain parameters for the LQR-PID controller. The strategy of the hybrid controller is based on the idea that the parameters of the PID controller are calculated using gain elements of the LQR optimal controller.

Chapter 3 presents the alternative optimized tuning approach of the PID controller. First, the PID controller parameters are optimized for the tracking response. Then, a simple disturbance estimator is introduced to substantially increase the disturbance-rejection performance. The proposed approach can still be used on the process data given either in parametric or nonparametric form. It will be shown how to achieve the best trade-off between performance and noise attenuation.

To solve the tuning problem in highly complex industrial processes, Chapter 4 concerns a controller adjustment method based on the internal product of PID terms. A propagation matrix (PM) is generated by the numerator coefficients of the plant transfer function (TF). In the proposed method, each term of the PID controller is influenced by each of the numerator and the denominator coefficients. Mathematical models of practical plants were employed to evaluate the proposed method. The obtained results demonstrated an assertive improvement in the adjustment gains from PID actions, thereby validating it as a promising alternative to conventional methods.

In Chapter 5, an application to the magnetic levitation (maglev) system is considered. To account for the complexity in the design procedure, this chapter presents a practical controller for the high-positioning performance of a magnetic levitation system. Three strategies of the proposed controller are the PI-PD controller is to enhance transient response, the model-based feedforward (FF) control is incorporated with the PI-PD controller to enhance the overshoot reduction characteristic in attaining a better transient response, and lastly, the disturbance compensator (Kz) is integrated as an additional feedback element to reduce the sensitivity function magnitude for robustness enhancement.

In Chapter 6, the estimation of multiparameters of complex systems based on the extended PID controllers is considered. Based on the results, with the introduction of a binary control mechanism, the integral item of the nonlinear PID controller could deal with the uncertain part of the complex system, which can also be called the new stripping principle (NSP), the new multiparameter estimation methods are given. Such kind of effort will improve the identification or cognition for certain kinds of complex systems, and it will also provide a completely different and effective way for the estimation methods in mathematical statistics.

Ever since the idea of publishing this book has been issued, we have received more than ten proposals and suggestions on the relevant chapters, given the current pandemic situation. We have accepted six of them to be included in this book. The intention of this book is to provide an active spur that will hopefully induce someone to come forward with his/her valuable contributions and to offer more relevant achievements in the future.

As this book is about to be published, I would like to express my deepest gratitude to all the authors of this book for their tremendous efforts in submitting their own chapters. I would like to give my sincere appreciation for the support from the IntechOpen publishing working team, and Miss Sara Debeuc in particular, for their kind support throughout the entire publishing process.

Wei Wang
School of Mathematics,
Renmin University of China,
Beijing, China

Chapter 1

Variable, Fractional-Order PID Controller Synthesis Novelty Method

Piotr Ostalczyk and Piotr Duch

Abstract

The novelty method of the discrete variable, fractional order PID controller is proposed. The PID controllers are known for years. Many tuning continuous time PID controller methods are invented. Due to different performance criteria there are optimized three parameters: proportional, integral and differentiation gains. In the fractional order PID controllers there are two additional parameters: fractional order integration and differentiation. In the variable, fractional order PID controller fractional orders are generalized to functions. Nowadays all PID controllers are realized by microcontrollers in a discrete time version. Hence, the order functions are discrete variable bounded ones. Such controllers offer better transient characteristics of the closed loop systems. The choice of the order functions is still the open problem. In this Section a novelty intuitive idea is proposed. As the order functions one applies two spline functions with bounded functions defined for every time subinterval. The main idea is that in the final time interval the variable, fractional order PID controller transforms itself to the classical one preserving the stability conditions and zero steady-state error signal. This means that in the last time interval the discrete integration order is -1 and differentiation is 1.

Keywords: fractional-Calculus, PID controller, discrete system

1. Introduction

A continuous-time proportional–integral–derivative controller (PID controller) [1] invented almost 100 years ago is one of the most widely applied controllers in the closed-loop systems [2] with many industrial applications [3–5]. Currently the continuous-time control is successively replaced by discrete-time one in which the integration is replaced by a summation and differentiation by a difference evaluation. So, in the discrete PID controller the classical integral is replaced by a sum and the derivative by a backward difference, [6]. The discrete controller's PID algorithm is mainly realized by micro-controllers [7].

At 70s of the 20-th Century the Fractional Calculus [8] with a great success started a considerable attention in mathematics and engineering [9–12]. Now, the fractional-order backward-difference (FOBD) and the fractional-order backward sum (FOBS) [6, 13] are applied in the dynamical system modeling [14] and discrete control algorithms. The continuous-time FOPID controllers are more difficult in a practical realization [15–18].

There are numerous continuous and discrete-time PID and FOPID controller synthesis methods [16, 19–31]. One should mention that the optimisation of the closed-loop system in this case is more complicated because of the controller optimization. Apart from the three classical gains there are two additional parameters, namely, a fractional order of differentiation and summation [32]. The FOPID control characterizes by slow achieving the steady state and growing calculation "tail" [12].

In the paper a novelty variable, the fractional-order PID (VFOPID) [6, 28, 33–41] controller synthesis is proposed. It consists of dividing the closed-loop system discrete-transient time division into the finite time intervals over which are defined fractional orders summation and differentiation functions. The main idea is that for the final infinite interval $[k_L, +\infty)$ the difference order equals 0 and the summation is $-!$ preserving quick reaching the zero steady state value. Thus, in the VFOPID control the disadvantages of FOPID are extracted. One should admit that in the FOPID or VFOPID control the microcontrollers are numerically loaded.

Fractional-orders systems are characterized by the so called system "memory". This, in practice, means that in every step the FOPID controller computes its output signals taking into account step-by-step linearly computed number of samples. This causes in practice the micro-controllers realization problems. It is known as "Finite memory principle" [12].

The paper is organized as follows. In Section 2 the basic information related to the fractional calculus and variable, fractional order Grünwald-Letnikov backward difference is given. The main result of the paper includes Section 3. It contains the proposed VFOPID controller synthesis method with the proposal of the order functions form. The brief description of the controller parameters evaluation algorithm is given. The investigations are supported by a numerical example presented in Chapter 4.

2. Mathematical preliminaries

In the paper the following notation will be used. $\mathbb{N}_0 = \{0, 1, 2, 3, \dots\}$, $\mathbb{N}_l = \{l, l+1, l+2, \dots\}$ $\mathbb{R}_+ = [0, +\infty)$. 0_k will denote the zero column vector of dimensions $(k+1) \times 1$ whereas $\mathbf{0}_{k,k}$ is $(k+1) \times (k+1)$ zero matrix. Similarly will be denoted a $(k+1) \times (k+1)$ unit matrix $\mathbf{1}_k$.

In general, a fractional-order functions will be denoted by Greek letters $\nu(\cdot) : \mathbb{N}_0 \to \mathbb{R}$ $\nu(\cdot)$ whereas the integer orders will be denoted by Latin ones $n \in \mathbb{R}_+$. In practice, for $l \in \mathbb{N}_0$: $0 < \nu(l) \leq 1$. For $k, l \in \mathbb{N}_0$ and a given order function $\nu(l)$ the function of two discrete variables $k, l \in \mathbb{N}_0$ is defined by the following formula: $a^{[\nu(l)]}(k)$ as follows:

Definition 2.1. For $k, l \in \mathbb{N}_0$ and a given order function $\nu(\cdot)$ one defines the coefficients function of two 13 discrete variables as

$$a^{[\nu(l)]}(k) = \begin{cases} 1 & \text{for} \quad k = 0 \\ (-1)^k \dfrac{\nu(l)(\nu(l)-1)\cdots(\nu(l)-k+1)}{k!} & \text{for} \quad k \in \mathbb{N}_1 \end{cases} \tag{1}$$

One should mention that function (1) for $\nu(l) = n(l) = \text{const} \in \mathbb{N}_0$

$$a^{[n]}(k) = \begin{cases} 1 & \text{for} \quad k = 0 \\ \dfrac{n(n-1)\cdots(n-k+1)}{(-1)^k!} & \text{for} \quad k \in [1, n] \\ 0 & \text{for} \quad k \in \mathbb{N}_{n+1} \end{cases} \tag{2}$$

The above function will be named as: the "oblivion function" or "decay function".

2.1 Variable, fractional-order backward difference

Next one defines the Grünwald–Letnikov variable, fractional-order backward difference (VFOBD). For a discrete-variable bounded real-valued function $f(\cdot)$ defined over a discrete interval $[0, k]$ the VFOBD is defined as a sum (see for instance [6, 9]).

Definition 2.2. The VFOBD with an order function ν, with values $\nu(k) \in [0, 1]$, is defined as a finite sum, provided that the series is convergent

$$
\begin{aligned}
{}_{k_0}\Delta_k^{\nu(k)} f(k) &= \sum_{i=0}^{k-k_0} a^{\nu(k)}(i) f(k-i) \\
&= \begin{bmatrix} 1 & a^{\nu(k)}(1) & a^{\nu(k)}(2) & \cdots & a^{\nu(k)}(k_0) \end{bmatrix} \begin{bmatrix} f(k) \\ f(k-1) \\ \vdots \\ f(k-k_0) \end{bmatrix}
\end{aligned} \tag{3}
$$

Relating to (2) as the first special case of the defined above VFOBD and a constant order function $\nu(k) = \nu = \text{const}$ from (2.1) one gets the fractional-order backward difference (FOBD). The second special case is for a constant integer order function $\nu(k) = \nu = n = \text{const}$ where the integer-order backward difference (IOBD) is a classical one.

Equality (3) is valid for $k, k-1, k-2, \dots, k_0+1, k_0$. Hence, one gets a finite set of equations. Collecting them in a vector matrix form one gets

$$
{}_{k_0}^{GL}\Delta_k^{[\nu(k)]} \mathbf{f}(k) = {}_{k_0}\mathbf{A}_k^{[\nu(k)]} \mathbf{f}(k), \tag{4}
$$

where

$$
{}_{k_0}\mathbf{A}_k^{[\nu(k)]} = \begin{bmatrix} 1 & a^{[\nu(k)]}(1) & \cdots & a^{[\nu(k)]}(k-k_0) \\ 0 & 1 & \cdots & a^{[\nu(k-1)]}(k-k_0-1) \\ \vdots & \vdots & & \vdots \\ 0 & 0 & \cdots & a^{[\nu(k_0+1)]}(1) \\ 0 & 0 & \cdots & 1 \end{bmatrix} \tag{5}
$$

$$
\mathbf{f}(k) = \begin{bmatrix} f(k) \\ f(k-1) \\ \vdots \\ f(k_0) \end{bmatrix}, \tag{6}
$$

$$
{}_{k_0}^{GL}\Delta_k^{[n(k)]} \mathbf{f}(k) = \begin{bmatrix} {}_{k_0}^{GL}\Delta_k^{[\nu(k)]} f(k) \\ \vdots \\ {}_{k_0}^{GL}\Delta_{k_0}^{[\nu(k_0)]} f(k_0) \end{bmatrix}.
$$

2.2 Variable, fractional-order linear time-invariant difference equations

On the base of the Grünwald-Letnikov variable, fractional-order linear time-invariant backward-difference the difference Eqs. (GL-VFOBE) for $i = 1, 2, \cdots, p$ and $j = 1, 2, \cdots, q$ representing discrete models of real dynamical systems or discrete control strategies are defined by the variable, fractional-order linear time-invariant difference equation (VFODE). $h > 0$ denotes the sampling time.

$$\sum_{l=0}^{n_i} a_{i,l} {}_{k_0}^{GL}\Delta_k^{[\nu_{i,l}(k)]} y(kh) = \sum_{l=0}^{m_i} b_{i,l} {}_{k_0}^{GL}\Delta_k^{[\mu_{i,l}(k)]} u(kh) \tag{7}$$

where $m_i \le n_i$, $\nu_{n_{i,l}}(k) \ge \nu_{n_{i,l-1}}(k) \ge \cdots \nu_{i,1}(k) \ge \nu_{i,0}(k) = 0$, $\mu_{m_{i,l}}(k) \ge \mu_{m_{i,l-1}-1}(k) \ge \cdots$ $\mu_{i,1}(k) \ge \mu_{i,0}(k) \ge 0$, $a_{i,l}$ and $b_{i,l}$ are constant coefficients for $l = 0, 1, \cdots, n_i$ and $l = 0, 1, \cdots, m_i$, respectively. It is assumed that $a_{0,n_0} = 1$.

According to the notation (5) Eq. (7) takes the form

$$\sum_{l=0}^{n_i} a_{i,lk_0} \mathbf{A}_k^{[\nu_{i,l}(k)]} \mathbf{y}(k) = \sum_{l=0}^{m_i} a_{i,lk_0} \mathbf{A}_k^{[\mu_{i,l}(k)]} \mathbf{u}(k) \tag{8}$$

The vector $\mathbf{u}_j(k)$ satisfies the condition $\mathbf{u}_j(k) = \mathbf{0}_k$ for $k < k_0$. In the general solution of (8) to the assumed $\mathbf{u}_j(k)$ and initial conditions vector $\mathbf{y}_{i,k_0-1} = \begin{bmatrix} y_{i,k_0-1} & y_{i,k_0-2} & \cdots \end{bmatrix}^{\mathrm{T}}$ (T denotes the transposition) must be taken into account with $-\infty = k_0' < 0 \le k_0 \le k$. Then, the infinite number of initial conditions (8) are formed in the following vector

$$\mathbf{y}_{i,k_0-1} = \begin{bmatrix} y_{i,k_0-1} \\ y_{i,k_0-2} \\ \vdots \end{bmatrix} \tag{9}$$

and the combined Eq. (8) is of the form

$$\begin{aligned} &\begin{bmatrix} \sum_{l=0}^{n_i} a_{ij,lk_0} \mathbf{A}_k^{[\nu_{i,l}(k)]} & \sum_{l=0}^{n_i} a_{i,l-\infty} \mathbf{A}_{k_0-1}^{[\nu_{i,l}(k)]} \end{bmatrix} \\ &\quad \times \begin{bmatrix} \mathbf{y}_i(kh) \\ \mathbf{y}_{i,k_0-1} \end{bmatrix} = \sum_{l=0}^{m_i} b_{ij,lk_0} \mathbf{A}_k^{[\mu_{i,l}(k)]} \mathbf{u}(kh) \end{aligned} \tag{10}$$

or after simple transformation

$$\begin{aligned} \sum_{l=0}^{n_i} a_{i,lk_0} \mathbf{A}_k^{[\nu_{i,l}(k)]} \mathbf{y}(kh) &= \sum_{l=0}^{m_i} b_{i,lk_0} \mathbf{A}_k^{[\mu_{i,l}(k)]} \mathbf{u}(kh) \\ &\quad - \sum_{l=0}^{n_i} a_{i,l-\infty} \mathbf{A}_{k_0-1}^{[\nu_{i,l}(k)]} \mathbf{y}_{i,k_0-1} \end{aligned} \tag{11}$$

2.3 Main assumptions

To preserve the VFOBDE order one assumes that

$$1 + \sum_{l=0}^{n_i-1} a_{i,l} \neq 0 \ \text{ for } i = 1, 2, \cdots, p \tag{12}$$

In the transfer functions defined by the one-sided **Z** transform one assumes zero initial conditions. Following this assumption equality (11) simplifies to

$$\sum_{l=0}^{n_i} a_{ik_0} \mathbf{A}_k^{[\nu_i(k)]} \mathbf{y}(k) = \sum_{l=0}^{m_i} b_{ik_0} \mathbf{A}_k^{[\mu_i(k)]} \mathbf{u}(k) \tag{13}$$

Defining matrices

$$_{\mathbf{k_0}}\mathbf{D}_{\mathbf{k}}^{[\nu_{\mathbf{P}}(\mathbf{k})]} = \sum_{l=0}^{n_i} a_{ik_0} \mathbf{A}_k^{[\nu_i(k)]} \tag{14}$$

$$_{\mathbf{k_0}}\mathbf{N}_{\mathbf{k}}^{[\mu_{\mathbf{P}}(\mathbf{k})]} = \sum_{l=0}^{m_i} b_{ik_0} \mathbf{A}_k^{[\mu_i(k)]} \tag{15}$$

one gets

$$_{k_0}D_k^{[\nu_P(k)]} \mathbf{y}(kh) = {}_{k_0}N_k^{[\mu_P(k)]} \mathbf{u}(kh) \tag{16}$$

Under assumption (12) $_{k_0}D_k^{[\nu(k)]}$ is invertible, so for $k_0 = 0$ one can write

$$\mathbf{y}(kh) = \left[{}_0 D^{[\nu_P(k)]_k}\right]^{-1} {}_0 N^{[\mu_P(k)]_k} \mathbf{u}(kh) \tag{17}$$

Denoting

$$_0G_k^{[\nu_P(k)]} = \left[{}_0D_k^{[\nu_P(k)]}\right] {}_0N_k^{[\mu(k)]} \tag{18}$$

one gets similar to the transfer function description

$$\mathbf{y}(kh) = {}_0G_k^{[\nu_P(k), \mu_P(k)]} \mathbf{u}(kh) \tag{19}$$

or for simplicity

$$\mathbf{G}_o(kh) = {}_0G_k^{[\nu_P(k), \mu_P(k)]} \tag{20}$$

Remark 2.1. Though the relation (19) looks similar to the classical discrete transfer function it is different by the real discrete variables. It relates discrete SISO systems by vectors and matrices related to its dimensions $k + 1 \in \mathbb{N}_0$.

2.4 VFO linear system description

One considers a closed-loop system illustrated in **Figure 1**. Where a plant is described by (19) where $e(kh)$ and $u(kh)$.

2.4.1 VFO_PID

The classical PID controller output is desribed by three terms

$$u(kh) = K_P e(kh) + K_{I0} \Delta_k^{-\mu(k)} e(kh) + K_{D0} \Delta_k^{\nu(k)} e(kh) \tag{21}$$

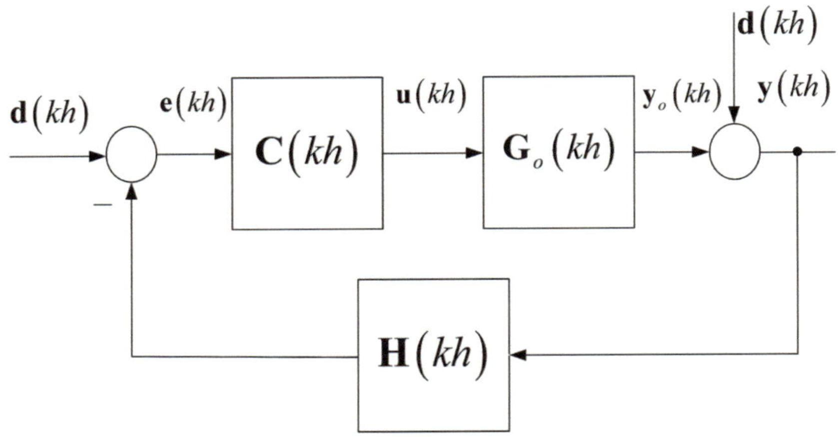

Figure 1.
Closed-loop system.

and in the convention proposed above as

$$\mathbf{u}(kh) = K_P \mathbf{1}_k \mathbf{e}(kh) + K_{D0} \mathbf{G}_k^{[\nu_C(k)]} \mathbf{e}(kh) + K_{I0} \mathbf{G}_k^{[-\mu_C(k)]} \mathbf{e}(kh) \tag{22}$$

which may be expressed as

$$\mathbf{u}(kh) = \left[K_P \mathbf{1}_k + K_{D0} \mathbf{G}_k^{[\nu_C(k)]} + K_{I0} \mathbf{G}_k^{[-\mu_C(k)]} \right] \mathbf{e}(kh) \tag{23}$$

where $\nu_C(k), \mu_C(k) \geq 0$ and controlling and error signals are denoted as $\mathbf{u}(k)$ and $\mathbf{e}(k)$, respectively. Then, denoting.

Remark 2.2. The plant may be described by classical integer order, fractional or even variable, fractional - order difference equations. The matrix - vector description used makes it possible.

$$\mathbf{C}(kh) = K_P \mathbf{1}_k + K_{D0} \mathbf{G}_k^{[\nu_C(k)]} + K_{I0} \mathbf{G}_k^{[-\mu_C(k)]} \tag{24}$$

one gets a VFOPID controller transfer function-like description

$$\mathbf{u}(kh) = \mathbf{C}(kh)\mathbf{e}(kh) \tag{25}$$

To simplify the description one assumes a sensor matrix as

$$\mathbf{H}(kh) = \mathbf{1}_k \tag{26}$$

The closed-loop system is presented in **Figure 1** from which one gets the following relations

$$\begin{aligned} \mathbf{y}(kh) = {} & [\mathbf{1}_k + \mathbf{G}_o(kh)\mathbf{C}(kh)\mathbf{H}(kh)]^{-1}\mathbf{G}_o(kh)\mathbf{C}(kh)\mathbf{r}(kh) \\ & + [\mathbf{1}_k + \mathbf{G}_o(kh)\mathbf{C}(kh)\mathbf{H}(kh)]^{-1}\mathbf{d}(kh) \end{aligned} \tag{27}$$

where

- $\mathbf{r}(kh)$ - a reference signal vector,
- $\mathbf{d}(kh)$ - an external disturbance signal vector,

- $\mathbf{y}_o(kh)$ - a plant output signal vector,
- $\mathbf{y}(kh)$ - a closed-loop system output signal vector,
- $\mathbf{e}(kh)$ - a closed-loop system error signal,

A system error is evaluated by the formula

$$\begin{aligned}\mathbf{e}(kh) = &[\mathbf{1}_k + \mathbf{G}_o(kh)\mathbf{C}(kh)\mathbf{H}(kh)]^{-1}\mathbf{r}(kh) \\ &-[\mathbf{1}_k + \mathbf{G}_o(kh)\mathbf{C}(kh)\mathbf{H}(kh)]^{-1}\mathbf{H}(kh)\mathbf{d}(kh)\end{aligned} \tag{28}$$

3. Variable, fractional-order PID controller synthesis

In the synthesis of the classical PID controller there are three parameters to evaluate. Namely, K,K_I,K_D known as the proportional, integral and differential gains. In the fractional-order PID controllers there are two additional parameters: the differentiation order $\nu(kh) \in \mathbb{R}_+$ and the integration one $-\mu(kh) \in \mathbb{R}_+$. In the variable, fractional-order PID controller the mentioned orders are generalized to functions. This means that there are three constant coefficients and two discrete variable functions to find

$$K_P, K_I, K_D, \nu(kh), \mu(kh) \tag{29}$$

In the rejection of the external disturbation one can assume that $\mathbf{r}(kh) = \mathbf{0}$ so Eq. (29) simplifies to

$$\mathbf{e}(kh) = -[\mathbf{1}_k + \mathbf{G}_o(kh)\mathbf{C}(kh)\mathbf{H}(kh)]^{-1}\mathbf{H}(kh)\mathbf{d}(kh) \tag{30}$$

Usually the sensor matrix $\mathbf{H}(kh)$ is treated as constant, by assumption that sensors do not introduce its own dynamics to the system. Hence, $\mathbf{H}(kh) = \mathbf{H} = \text{const}$. It may be assumed that $\mathbf{H} = h_0\mathbf{1}_k$ or further, for $h_0 = 1$, formula (30) takes a form

$$\mathbf{e}(kh) = -[\mathbf{1}_k + \mathbf{G}_o(kh)\mathbf{C}(kh)]^{-1}\mathbf{d}(kh) \tag{31}$$

The optimal parameters (29) are evaluated due to the assumed optymality criterion. The most popular is so called ISE one (Integral of the Squared Error) or in the discrete-system case: Sum of the Squared Error (SSE).

$$SSE[K_P, K_I, K_D, \nu(kh), \mu(kh)] = \sum_{i=0}^{k_max} e(ih)^2 h = \mathbf{e}(kh)^{\mathrm{T}}\mathbf{e}(kh)h \tag{32}$$

Substitution of (31) into (32) gives

$$\begin{aligned}&SSE[K_P, K_I, K_D, \nu(kh), \mu(kh)] \\ &= \mathbf{d}(kh)^T[\mathbf{1}_k + \mathbf{G}_o(kh)\mathbf{C}(kh)]^{-T}[\mathbf{1}_k + \mathbf{G}_o(kh)\mathbf{C}(kh)]^{-1}\mathbf{d}(kh)\end{aligned} \tag{33}$$

In the proposed VFOPID controller synthesis method with partially intuitive and supported by closed-loop systems synthesis experience the classical optimisation due to the performance criterion (32) is performed. The pre-defined differentiation and integration order functions orders are as follows

$$\nu(kh) \geq 0 \tag{34}$$

$$\nu(kh) = \begin{cases} \nu_1(kh) & \text{for} \quad k \in [0, k_{N1}) \\ \nu_2(kh) & \text{for} \quad k \in [k_{N1}, k_{N2}) \\ & \vdots \\ \nu_N(kh) & \text{for} \quad k \in [k_{NN-1}, k_N N) \\ 0 & \text{for} \quad k \in [k_{NN}, +\infty) \end{cases} \tag{35}$$

and

$$\mu(kh) \leq 0 \tag{36}$$

$$\mu(kh) = \begin{cases} \mu_1(kh) & \text{for} \quad k \in [0, k_{M1}) \\ \mu_2(kh) & \text{for} \quad k \in [k_{M1}, k_{M2}) \\ & \vdots \\ \mu_N(kh) & \text{for} \quad k \in [k_{MM-1}, k_M M) \\ -1 & \text{for} \quad k \in [k_{MM}, +\infty) \end{cases} \tag{37}$$

Every function $\nu_i(kh)$ for $i = 1, 2, \cdots, N$ and $\mu_i(kh)$ for $i = 1, 2, \cdots, M$ is characterized by a sets of parameters c_{ij} and d_{ij}, respectively.

In the classical closed-loop system with PID controller there is introduced the integration part preserving the steady - state error signal tending to zero. So, in (38) there is a constant order -1 for $k \geq [K_{MM}, +\infty)$.

Now, for initially assumed order functions one applies the following algorithm based on well known Gauss method.

1. Chose a starting set of coefficients $K_P, K - I, K_D, c_{11}, \cdots$ and $d_{11}, \cdots$,
2. Applying the classical Gauss algorithm find a minimal SSE performance index value alongside the first variable (eg. K_P),
3. Repeat step 2 for the next parameter,
4. If the SSE value is satisfactory stop else return to step 2.

Remark 3.1. Algorithm described above can be applied also to the classical discrete PID controller with three parameters.

4. Numerical example

One considers a closed-loop system depicted in **Figure 1**.

A plant is described by a transfer function

$$G_o(s) = \frac{b_0}{s^2 + a_1 s + a_0} \tag{38}$$

where

- $a_1 = 0.5$
- $a_0 = 0.1$
- $b_0 = a_0$

The plant is discretized with the sampling time $h = 0.5$ and a VFOPID controller is applied

$$\nu(kh) = \begin{cases} \nu_1(kh) = 1 & \text{for} \quad k = 0 \\ 0 & \text{for} \quad k \in [1, +\infty) \end{cases} \tag{39}$$

$$\mu(kh) = \begin{cases} \mu_1(kh) = -1 + d_1 e^{d_2(kh-1h)} & \text{for} \quad k = [0, 10] \\ -1 & \text{for} \quad k \in (10, +\infty) \end{cases} \tag{40}$$

and controller gains K_P, k_I, K_D and order function parameters d_1, d_2.

Hence, there are 5 parameters to evaluate. Due to the performance index (33) the optimal parameters are as follows

- $K_P = 1.000$
- $K_i = 0.514$
- $K_D = 0.890$

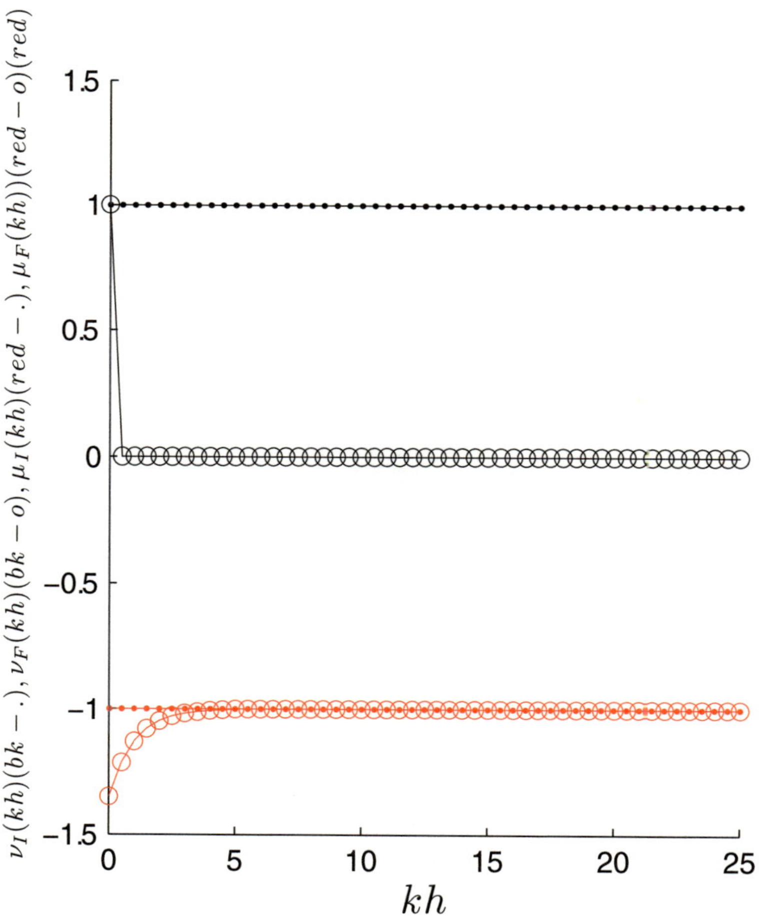

Figure 2.
VFOPID controller order functions: $\nu(kh)$ (in black) and $\mu(kh)$ (in red).

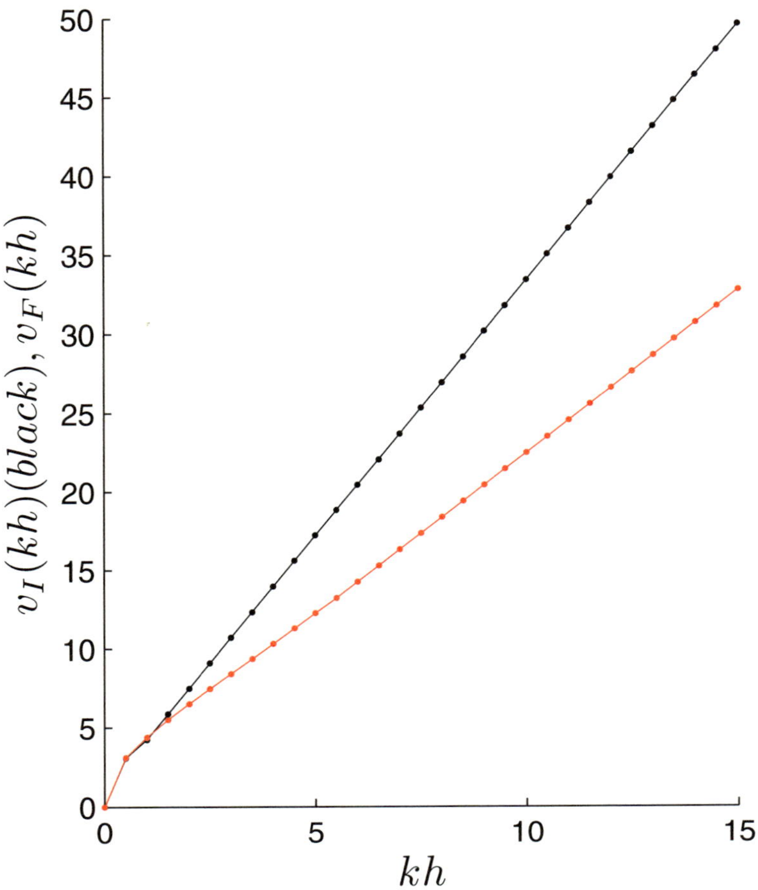

Figure 3.
VFOPID (in red) and IOPID (in black) controller unit step response.

- $d_1 = -0.35$
- $d_2 = -0.5$

The VFOPID controller order functions are plotted in **Figure 2** whereas the PID and VFOPID controllers unite step responses are given in **Figure 3**.

The achieved VFOPID controller synthesis result is compared with the classical discrete-time PID controller optimized due to criterion (30). The optimal parameters are

- $K_P = 1.00$
- $K_i = 0.81$
- $K_D = 0.90$

Figure 4 contains the closed - loop systems with PID (in blue) and VFOPID (in red) controllers unit step responses. There is included a plant unit step response of the plant (in black.)

In **Figure 5** the controlling signals are presented (PID - in black, VFOPID - in red). The controlling signals have typical shapes: first differentiation action and

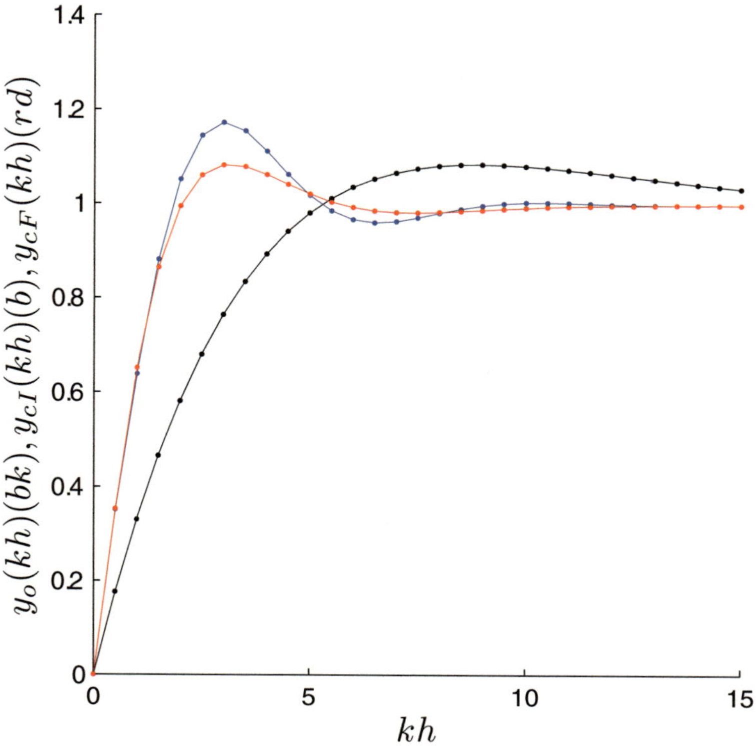

Figure 4.
The closed-loop system response with the VFOPIS (in red) and IOPID controllers (in blue).

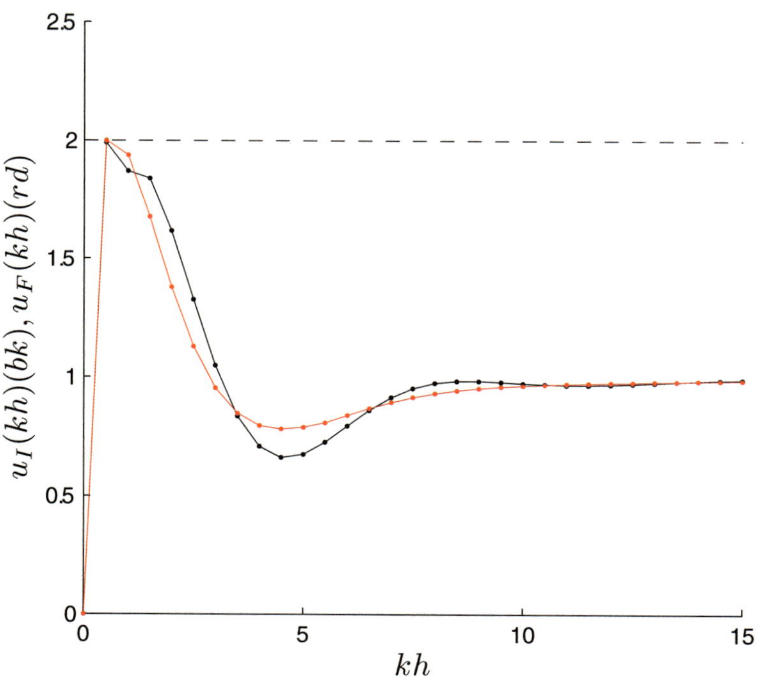

Figure 5.
The closed-loop controlling signals.

finally the classical integration preserving zero steady - state closed - loop system error.

Remark 4.1. In the Numerical example proposed here the VFOPID and the classical PID controllers maximal control signal values are the same reaching assumed bounding value $\max[u_I(kl)]$, $\max[u_F(kh)] = 2$.

Remark 4.2. In the Numerical example

$$\begin{aligned} SSE[K_P, K_I, K_D, 1, -1)] &= 1.3312 \\ SSE[K_P, K_I, K_D, \nu(kh), \mu(kh)] &= 1.2899 \end{aligned} \tag{41}$$

5. Conclusions

One should emphasize that the proposed solution of the VFOPID controller do not guarantee the absolute optimum of the closed-loop control system synthesis. It proves that the proposal of a physically realizable VFOPID controller by micro-controller (with finite memory) leads to better results due to the assumed performance criterion.

The main idea of the proposed method is to assume *a priori* the order functions with unknown parameters. In the VFOPID controller synthesis essential is an assumption that the summation order equals 1 One can express the action as the assumption of skeleton order functions with unknown parameters evaluated further in an SSE optimization algorithm.

Here, it is worth mentioning that there are still open problems of the VFOPID controllers tuning.

- One should define a program evaluating the order functions.
- For evaluation of the VFOPID controller parameters one can apply another optimization methods. It seems that optimization methods based on the artificial intelligence will be very effective.
- Another performance index may be applied. Some penalty functions may be introduced to SSE as well a term taking into account the minimal value of the error signal.

Acknowledgments

The work was supported by the National Science Center Poland by Grant no. 2016/23/B/ ST7/03686.

Abbreviations

PID	proportional-integral-derivative controller
FOBD	fractional-order backward difference
FOBS	fractional-order backward sum
FOPID	fractional-Order proportional, fractional-order integral and differential controller
VFOPID	variable, fractional-order PID controller
SSE	squared sum of the error

Author details

Piotr Ostalczyk[1*†] and Piotr Duch[2†]

1 Institute of Applied Computer Sciences, Lodz University of Technology No. 1, Lodz, Poland

2 Institute of Applied Computer Sciences, Lodz University of Technology No. 2, Lodz, Poland

*Address all correspondence to: postalcz@p.lodz.pl

† These authors contributed equally.

References

[1] Åström, K. J. (1995). PID controllers: theory, design and tuning. In Instrument Society of America. https://doi.org/1556175167

[2] Kailath T. (1980) Linear Systems. Englewood Cliffs, N.J.: Prentice-Hall, Inc. ISBN 0–13–536961–4

[3] Chopade, A. S., Jonathan, L., Junghare, A. S., Aware, M. V. (2014). Hardware in Loop, real time implementation of Fractional Order PID controller on Magnetic Levitation System. In Symposium on Advances in Control & Instrumentation (SACI-2014). Vol.24, pp.26. DOI: 10.13140/2.1.3151.9521

[4] Idir, A., Kidouche, M., Bensafia, Y., Khettab, K., & Tadjer, S. A. (2018). Speed control of DC motor using PID and FOPID controllers based on differential evolution and PSO. International Journal of Intelligent Engineering and Systems, 11(4), 241–249. https://doi.org/10.22266/ijies2018.0831.24

[5] Mirghasemi, S. A., Necsulescu, D., Sasiadek, J. (2019). Quadcopter Fractional Controller Accounting for Ground Effect. Proceedings of the 12th International Workshop on Robot Motion and Control, (2), 32–37. https://doi.org/10.1109/romoco.2019.8787366

[6] Ostalczyk P.: Discrete Fractional Calculus. Selected Applications. Applications in control and Image processing Series in computer Vision - vol. 4, New Jersey: World Scientific; 2016, 361 p.: 745–750. ISBN: 978–981–4725-66-8

[7] Matusiak, M., & Ostalczyk, P. (2019). Problems in solving fractional differential equations in a microcontroller implementation of an FOPID controller. Archives of Electrical Engineering, 68(3), 565–577. https://doi.org/10.24425/aee.2019.129342

[8] Oustaloup A. (1995) La dérivation non entière: théorie, synthèse et applications. Hermes. DOI 2–7462–2926-9

[9] Coimbra CF.M, (2003). Mechanics with variable-order differential operators, Ann. Phys., Leipzig, No. 11–12, pp. 692–703. DOI 10.1002/andp.200310032.

[10] Kilbas A.A, Srivastava H.M., Trujillo J. (2006) Theory and Applications of Fractional Differential Equations. Amsterdam. Elsevier. ISBN-13 978–0–444-51832-3

[11] Munoz-Hernandez, G. A., Mino-aguilar, G., Guerrero-castellanos, J. F., Peralta-sanchez, E. (2020). Fractional order PI-based control applied to the traction system of an electric vehicle (EV). Applied Sciences (Switzerland), 10(1). https://doi.org/10.3390/app10010364

[12] Podlubny, I. (1999). Fractional-order systems and PIλDμ-controllers. IEEE Transactions on Automatic Control, vol.44. No.1, 208–214. https://doi.org/10.1109/9.739144

[13] Baleanu D., Diethelm K., Trujillo J.J. (2012). Fractional Calculus: Models and Numerical Methods. Series on Complexity, Nonlinearity and Chaos. Singapore: World Scientific.

[14] Caponetto R., Dongola G., Fortuna G. and Petras I. (2010). Fractional Order Systems: Modeling and Control Applications. Singapore: World Scientific. ASIN : B00ADHLAM2.

[15] Mahmood, A. K., Abdul, S., & Saleh, R. (2018). Realization of fractional-order proportional-integral-derivative controller using fractance circuit. JEA JOURNAL OF ELECTRICAL ENGINEERING, Vol.2. No.1, pp.22–32. https://doi.org/10.1177/0959651818790809

[16] Monje, C. A., Vinagre, B. M., Chen, Y., Feliu, V., Lanusse, P., Sabatier, J. (2004). Proposals for fractional order PIλDμ tuning. Proceedings of the First IFAC Workshop on Fractional Differentiation and Applications, Bordeaux, France.

[17] Malesza, W., Sierociuk, D. (2018). Fractional variable order anti-windup control strategy. Bulletin of the Polish Academy of Sciences: Technical Sciences, Vol.66. No4. pp.427–432. DOI 10.24425/124258

[18] Ostalczyk, P., Rybicki T. (2008). Variable-fractional-order dead-beat control of an electromagnetic servo. Journal of Vibration and Control, 14(9–10), 1457–1471. https://doi.org/10.3182/20060719-3-PT-4902.00017

[19] Baleanu D., Machado J.T. Luo A.J. (2014), Fractional Dynamics and Control. New York: Springer-Verlag. ISBN 978–1-4-4614-0456-9

[20] Barbosa R.S., Silva M.F., Machado J. A.T: Tuning and Application of Integer and Fractional Order PID Controllers. (2009). Eds. Tenreiro Machado J.J., Pátani, Rudas L.J. Intelligent Engineering Systems and Computational Dybernetics. pp. 245–255-4. DOI: 10.1007/978-1-4020-8678-6_21

[21] Brzeziński, D. W., Ostalczyk, P. (2012). The Grunwald-Letnikov formula and its equivalent Horner's form accuracy comparison and evaluation for application to fractional order PID controllers. 2012 17th International Conference on Methods and Models in Automation and Robotics MMAR https://doi.org/10.1109/MMAR.2012.

[22] Dastjerdi, A. A., Vinagre, B. M., Chen, Y., HosseinNia, S. H. (2019). Linear fractional order controllers; A survey in the frequency domain. Annual Reviews in Control, 47, 51–70. https://doi.org/10.1016/j.arcontrol.2019.

[23] Das, S., Saha, S., Das, S.,Gupta, A. (2011). On the selection of tuning methodology of FOPID controllers for the control of higher order processes. ISA transactions, Vol.5. No.3, pp.376–388. DOI: 10.1016/j.isatra.2011.02.003

[24] De Keyser, R., Muresan, C. I., Ionescu, C. M. (2018). Autotuning of a robust fractional order PID controller. IFAC-PapersOnLine, 51(25), pp.466–471. https://doi.org/10.1177/0959651818790809

[25] El-Khazali, R. (2013). Fractional-order PIλDμ controller design. Computers & Mathematics with Applications, 66(5), 639–646. https://doi.org/10.1016/j.camwa.2013.02.015

[26] Gligor, A., & Dulău, T. M. (2017). Fractional Order Controllers Versus Integer Order Controllers. Procedia Engineering, 181, 538–545. https://doi.org/10.1016/j.proeng.2017.02.431

[27] Merrikh-Bayat, F., Mirebrahimi, N., Khalili, M. R. (2015). Discrete-time fractional-order PID controller: Definition, tuning, digital realization and some applications. International Journal of Control, Automation and Systems, 13(1), 81–90. https://doi.org/10.1007/s12555-013-0335-y

[28] Xue D., Liu L., Pan F. (2015, June). Variable-order fuzzy fractional PID controllers for networked control systems. ISA Transactions. Vo. 55. pp. 227–233. https://doi.org/10.1016/j.isatra.2014.09.012

[29] Oprzedkiewicz, K., Dziedzic, K. (2017). A Tuning of a Fractional Order PID Controller with the Use of Particle Swarm Optimization Method. https://doi.org/10.1007/978-3-319-59063-9_35

[30] Li, X., Wang, Y., Li, N., Han, M., Tang, Y., & Liu, F. (2017). Optimal fractional order PID controller design for automatic voltage regulator system based on reference model using particle

swarm optimization. International Journal of Machine Learning and Cybernetics, 8(5), 1595–1605. https://doi.org/10.1007/s13042-016-0530-2

[31] Merrikh-Bayat, F. (2012). General rules for optimal tuning the PIλDμ controllers with application to first-order plus time delay processes. Canadian Journal of Chemical Engineering. https://doi.org/10.1002/cjce.21643

[32] Liu, L., Pan, F., & Xue, D. (2015). Variable-order fuzzy fractional PID controller. ISA Transactions. https://doi.org/10.1016/j.isatra.2014.09.012

[33] Aydogdu, O., Korkmaz, M. (2019). Optimal Design of a Variable Coefficient Fractional Order PID Controller by using Heuristic Optimization Algorithms. International Journal of Advanced Computer Science and Applications, 10(3). https://doi.org/10.14569/IJACSA.2019.

[34] Chen, L., Chen, G., Wu, R., Lopes, A. M., Machado, J. A. T., Niu, H. (2020). Variable coefficient fractional-order PID controller and its application to a SEPIC device. IET Control Theory & Applications, 14(6), pp.900–908.

[35] Dabiri, A., Moghaddam, B. P., Machado, J. T. (2018). Optimal variable-order fractional PID controllers for dynamical systems. Journal of Computational and Applied Mathematics, 339, pp.40–48.DOI:10.1016/j.cam.2018.02.029Corpus ID: 19098724

[36] Ostalczyk, P. (2012). Variable-, fractional-order discrete PID controllers. 2012 17th International Conference on Methods and Models in Automation and Robotics, MMAR 2012, 534–539. https://doi.org/10.1109/MMAR.2012.6347829

[37] Ostalczyk, P., Brzeziński, D. W., Duch, P., Łaski, M., & Sankowski, D. (2013). The variable, fractional-order discrete-time PD controller in the IISv1.3 robot arm control. Open Physics (Central European Journal of Physics), 11(6), 750–759. https://doi.org/10.2478/s11534-013-0254-9

[38] Ostalczyk, P. (2019). Analysis of a closed loop system with DC micro-motor electrical drive with propeller and variable-fractional order PID controller. In 2019 20th International Carpathian Control Conference (ICCC). pp. 1–6. IEEE. ISBN 978–1–7281-0703-5

[39] Oziablo, P., Mozyrska, D., Wyrwas, M. (2020). Discrete-Time Fractional, Variable-Order PID Controller for a Plant With Delay. Entropy, Vol.22. No.7, 771. DOI:10.3390/e22070771

[40] Sierociuk, D., & Macias, M. (2013). Comparison of variable fractional order PID controller for different types of variable order derivatives. Proceedings of the 2013 14th International Carpathian Control Conference, ICCC 2013. https://doi.org/10.1109/CarpathianCC.2013.6560565

[41] Sierociuk D., Wiraszka M.S. (2018) A new Variable Fractional-Order PI Algorithm. IFAC PapersOnLine, 2018, vol.51. No.4. pp.745–750. DOI:https://doi.org/10.1016/j.ifacol.2018.06.207

Chapter 2

A Hybrid Control Approach Based on the Combination of PID Control with LQR Optimal Control

Ibrahim K. Mohammed

Abstract

Proportional Integral Derivative (PID) is the most popular controller that is commonly used in wide industrial applications due to its simplicity to realize and performance characteristics. This technique can be successfully applied to control the behavior of single-input single-output (SISO) systems. Extending the using of PID controller for complex dynamical systems has attracted the attention of control engineers. In the last decade, hybrid control strategies are developed by researchers using conventional PID controllers with other controller techniques such as Linear Quadratic Regulator (LQR) controllers. The strategy of the hybrid controller is based on the idea that the parameters of the PID controller are calculated using gain elements of LQR optimal controller. This chapter focuses on design and simulation a hybrid LQR-PID controller used to stabilize elevation, pitch and travel axes of helicopter system. An improvement in the performance of the hybrid LQR-PID controller is achieved by using Genetic Algorithm (GA) which, is adopted to obtain best values of gain parameters for LQR-PID controller.

Keywords: proportional integral derivative (PID), fractional order proportional integral derivative (FOPID), linear quadratic regulator (LQR), hybrid control system, genetic algorithm (GA)

1. Introduction

PID is regarded as the standard control structure of classical control theory. PID controllers are used successfully for single-input single-output (SISO) and linear systems due to their good performance and can be easily implemented. The control of complex dynamic systems using classic PID controllers is considered as a big challenge, where the stabilization of these systems requires applying a more robust controller technique. Many studies have proposed to develop a new hybrid PID controller with ability to provide better and more robust system performance in terms of transient and steady-state responses over the standard PID controllers. Lotfollahzade et al. [1] proposed a new LQR-PID controller to obtain an optimal load sharing of an electrical grid. The presented hybrid controller is optimized by Particle Swarm Optimization (PSO) to compute the gain parameters of the PID controller. A new hybrid control algorithm was presented by Lindiya et al. for power converters [2]. They adopted a conventional multi-variable PID and LQR algorithm for reducing cross-regulation in DC-to-DC converters. Sen et al.

IntechOpen

introduced a hybrid LQR-PID controller to regulate and monitor the locomotion of a quadruped robot. The gain parameters of the hybrid controller is tuned using the Grey-Wolf Optimizer (GWO) [3]. In [4] a new PID and LQR control system was proposed to improve a nonlinear quarter car suspension system.

The intent of this study is to design a new hybrid PID controller based on an optimal LQR state feedback controller for stabilization of 3DOF helicopter system. To this end an improvement in the system performance has been achieved in both the transient and steady-state responses. In the proposed system the classical PID and optimal LQR controller have been combined to formulate a hybrid controller system. Simulations were implemented utilizing Matlab programming environment to verify the efficiency and effectiveness of the proposed hybrid control method.

2. Controller theory

In this section, basics and theory of integer and fractional order PID controllers are presented. Theory of an intelligent LQR controller, which is used with PID controller to combine a hybrid control system, is also introduced.

2.1 Calssical PID controller

A PID is the most popular controller technique that is widely used in industrial applications due to the simplicity of its structure and can be realized easily for various control problems as the gain parameters of the controller are relatively independent [5, 6]. Basically, the controller provides control command signals $u(t)$ based on the error $e(t)$ between the demand input and the actual output of the system. The continuous time structure of the classical PID controller is as follows:

$$u(t) = K_p e(t) + K_i \int_0^t e(\tau)d\tau + K_d \frac{de(t)}{dt} \tag{1}$$

where K_p, K_i and K_d are the proportional, integral and differential components of the controller gain. These controller gain parameters should be tuned properly to enable the output states of the system to efficiently follow the desired input.

2.2 FOPID controller

FOPID is a special category of PID controller with fractional order derivatives and integrals. Its concept was introduced by Podlubany in 1997. During the last decade, this controller approach has attracted the attention of control engineers in both academic and industrial fields. Compared with the classical PID controller, it offers flexibility in dynamic systems design and more robustness.

2.2.1 Fractional order calculus

Fractional order calculus is an environment of calculus that generates the derivatives or integrals of problem functions to non-integer (fractional) order. This fractional order mathematical operation allows to establish a more accurate and concise model than the classical integer-order method. Moreover, the fractional order calculus can also produce an effective tool for describing dynamic behavior for control systems [7].

$$aD_t^{\alpha} = \begin{cases} \dfrac{d^{\alpha}}{dt^{\alpha}} & \alpha > 0 \\ 1 & \alpha = 1 \\ \int\limits_{\alpha}^{t} (dt)^{-\alpha} & \alpha < 0 \end{cases} \tag{2}$$

Fractional order calculus is a generalization of differentiation and integration to non-integer order fundamental operator which is denoted by aD_t^{α} where a and t are the operation limits and $\alpha(\alpha \in \mathcal{R})$ is the order of the operation. The formula of continuous differ-integral operator (aD_t^{α}) is defined as in Eq. (2) [8]. There are two commonly used definitions for general fractional differ-integral (aD_t^{α}), which are used for realization of control problem algorithm:

Grunwald – Letnikov (GL) definition:

$$aD_t^{\alpha} f(t) = \frac{d^{\alpha} f(t)}{dt^{\alpha}} = \lim_{h \to 0} h^{-\alpha} \sum_{j=0}^{[x]} (-1)^j \binom{\alpha}{j} f(t - jh) \tag{3}$$

where $[x]$ is integer part of x, $x = \frac{t-a}{h}$, h is time step and $\binom{\alpha}{j}$ is binomial coefficients, its expression is given by:

$$\binom{\alpha}{j} = \frac{\alpha(\alpha-1) \dots \dots ..(\alpha - j + 1)}{j!} \tag{4}$$

Riemann-Liouville definition:

$$aD_t^{\alpha} f(t) = \frac{1}{\Gamma(n-\alpha)} \frac{d^n}{dt^n} \int\limits_{a}^{t} \frac{f(\tau)}{(t-\tau)^{\alpha-n+1}} d\tau \tag{5}$$

where $n \in \mathbb{R}^{+}$. The condition for above equation is $n - 1 < \alpha < n, \Gamma(.)$ is called Gamma function, which its defination is given by:

$$\Gamma(X) = \int\limits_{0}^{\infty} z^{X-1} e^{-z} dz \tag{6}$$

Laplace transform of differ-integral operator $(\mathrm{aD}_{\mathrm{t}}^{\alpha})$ is given by expected form:

$$L\left[aD_t^{\alpha} f(t)\right] = \int_0^{\infty} e^{-st} aD_t^{\alpha} f(t) dt \tag{7}$$

$$L\left[aD_t^{\alpha} f(t)\right] = s^{\alpha} F(s) - \sum_{m=0}^{n-1} s(-1)^j 0D_t^{\alpha-m-1} f(t) \tag{8}$$

Where $F(s) = L\{f(t)\}$ is the normal Laplace transformation and n is an integer number that satisfies $n - 1 < \alpha \leq n$ and $s = jw$ denotes the Laplace transform variable.

2.2.2 Fractional order controller

Fractional order PID controller denoted by $\mathrm{PI}^{\lambda}\mathrm{D}^{\mu}$ was proposed by Igor Podlubny [9] in 1997. It is an extension of traditional PID controller where λ and μ

have non-integer fractional values. **Figure 1** shows the block diagram of the fractional order PID controller. The integer-differential equation defining the control action of a fractional order PID controller is given by:

$$u(t) = K_p e(t) + K_i D^{-\lambda} e(t) + K_d D^{-\mu} e(t) \quad (9)$$

Based on the above equation, it can be expected that the FOPID controller can enhance the performance of the control system due to more tuning knobs introduced. Taking the Laplace transform of Eq. (9), the system transfer function of the FOPID controller is given by:

$$G_{FOPID}(s) = K_p + \frac{K_i}{s^\lambda} + K_d s^\mu \quad (10)$$

Where λ and μ are arbitrary real numbers. Taking $\lambda = 1$ and $\mu = 1$ a classical PID controller is obtained. Thus, FOPID controller generalizes the classical PID controller and expands it from point to plane as shown in **Figure 2(b)**. This expansion provides the designer much more flexibility in designing PID controller and gives an opportunity to better adjust the dynamics of the control system. This increases robustness to the system and makes it more stable [10]. A number of optimization techniques can be implemented for getting the best values of the gain parameters of the controller.

2.3 LQR controller

Linear quadratic regulator is a common optimal control technique, which has been widely utilized in various manipulating systems due to its high precision in movement applications [11]. This technique seeks basically a tradoff betwwen a stable performance and acceptable control input [12]. Using the LQR controller in the design control system requires all the plant states to be measurable as it bases on

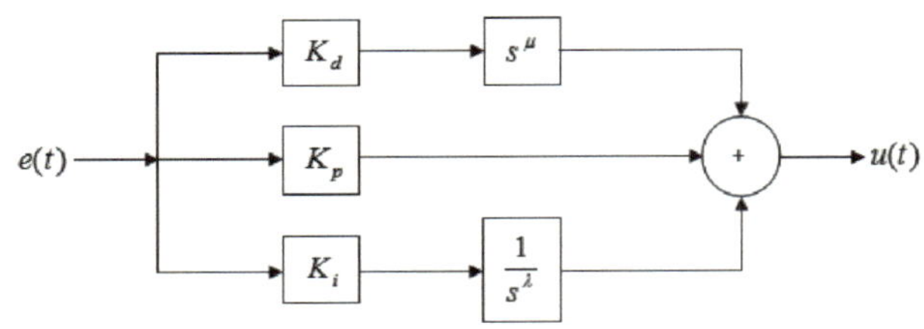

Figure 1.
The block diagram of a FOPID structure.

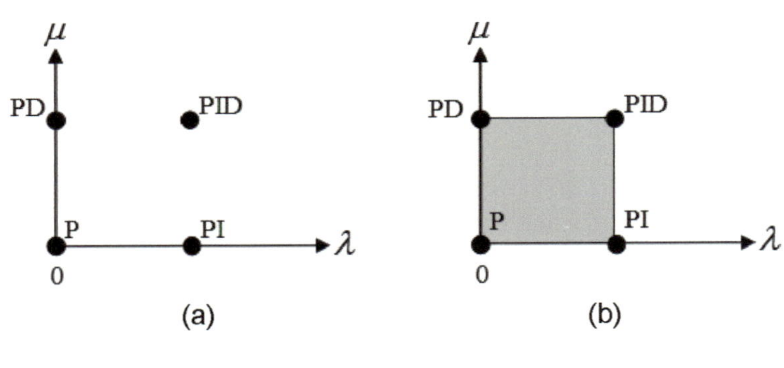

Figure 2.
PID controllers with fraction orders. (a) Classical. (b) Fractional order.

the full state feedback concept. Therefore, using the LQR controller to stabilize the 3DOF helicopter system based on the assumption that the system states are considered measurable. LQR approach includes applying the optimal control effort:

$$u(t) = -Kx(t) \tag{11}$$

Where K is the state feedback gain matrix, that will enable the output states of the system to follow the trajectories of reference input, while minimizing the following the cost function:

$$J = \int_0^\infty \left(x^T(t)Qx(t) - u^T(t)Ru(t)\right)dt \tag{12}$$

Where Q and R are referred to as weighting state and control matrices. The controller feedback gain matrix can be determined by using below equation:

$$K = R^{-1}B^TP \tag{13}$$

Where P is *(nxn)* matrix deterrmined from the solution of the following Riccati matrix equation:

$$A^TP + PA - PBR^{-1}B^TP + Q = 0 \tag{14}$$

For n^{th} order system with m^{th} input, the gain matrix and control input are given by:

$$\mathrm{K} = \begin{bmatrix} k_{11} & k_{12} & k_{13} & \dots & k_{1n} \\ k_{21} & k_{22} & k_{23} & \dots & k_{2n} \\ . & . & . & . & . \\ k_{m1} & k_{m2} & k_{m3} & \dots & k_{mn} \end{bmatrix} \quad \text{and} \quad u(t) = \begin{bmatrix} u_1 \\ u_2 \\ u_3 \\ . \\ u_m \end{bmatrix}$$

Based on the above expression, the control effort $u(t)$ of the system stated in Eq. (11) can be written as in Eq. (15). For the purpose of simplicity of control problem the weighting matrices Q and R are chosen as the diagonal matrices:

$$Q = blkdig(q_{11}, q_{22}, q_{33}, \dots \dots \dots, q_{nn}), R = blkdig(r_{11}, r_{22}, r_{33}, \dots, r_{mm})$$

so that the cost function Eq. (12) can be reformulated as in Eq. (16).

$$u(t) = -\begin{bmatrix} k_{11} & k_{12} & k_{13} & \dots & k_{1n} \\ k_{21} & k_{22} & k_{23} & \dots & k_{2n} \\ . & . & . & . & . \\ k_{m1} & k_{m2} & k_{m3} & \dots & k_{mn} \end{bmatrix} \begin{bmatrix} x_1 \\ x_2 \\ x_3 \\ . \\ xn \end{bmatrix} \tag{15}$$

$$J = \int_0^\infty \left(q_{11}{x_1}^2 + q_{11}{x_2}^2 + \dots.. + q_{nn}{x_n}^2 + r_{11}{u_1}^2 + r_{22}{u_2}^2 + \dots.. + r_{mm}{u_m}^2\right)dt \tag{16}$$

Where $q_{11}, q_{22}, q_{33}, \dots.., q_{nn}$ and $r_{11}, r_{22}, r_{33}, \dots.., r_{mm}$ denote the weighting elements of Q and R matrices respectively. The optimal control approach LQR is highly recommended for stabilizing complex dynamic systems as it basically looks for a compromise between the best control performance and minimum control input effort. Based on the LQR controller an optimum tracking performance can be investigated by a proper setting of the feedback gain matrix. To achieve this, the LQR controller is optimised by using GA tuning method which is adopted to obtain optimum elements values for of Q and R weighting matrices.

3. Tuning method

In this study, GA tuning approach has been invoked to tune the gain matrix of LQR controller used to approximate the gain parameters of PID controller for 3DOF helicopter system. GA is a global search optimization technique bases on the strategy of natural selection. This optimization method is utilized to obtain an optimum global solution for more control and manipulating problems. The procedure of GA approach includes three basic steps: selection, crossover and mutation, that constitute the main core of GA with powerful searching ability.

Selection: This step includes choosing individual genomes with high adaptive value from the current population to create mating pool. At present, there mainly are: sequencing choice, adaptive value proportional choice, tournament choice and so on. In order to avoid the best individuals of current population missing in the next generation due to destruction influence of crossover and mutation or selection error, De Jong put forward to the cream choice strategy [3xxx];

Crossover: This operation is the process of mimicking gene recombination of natural sexual reproduction, through combining the genetic information of two gens to create a new offspring contining more complicated gene structur. Reproduction may proceed in three stages as follows: (1) two newly reproduced strings are randomly selected from a Mating Pool; (2) a number of crossover positions along each string are uniformly selected at random and (3) two new strings are created and copied to the next generation by swapping string characters between the crossover positions defined before.

Mutation: In this process one or more indivisual values in a chromosome are altered from its initial state. This can result in entirely new gene values being added to the gene pool. This stage is also important by the view of preventing the genes local optimal points.

Applying these main operations creates new individuals which could be better than their parents. Based on the requirements of desired response, the sequence of GA optimization technique is repeated for many iterations and finally stops at generating optimum solution elements for the application problems. The sequence of the GA tuning method is presented in **Figure 3** [13, 14]. The steps of the GA loop are defined as follows:

1. Initial set of population.

2. Choose individuals for mating.

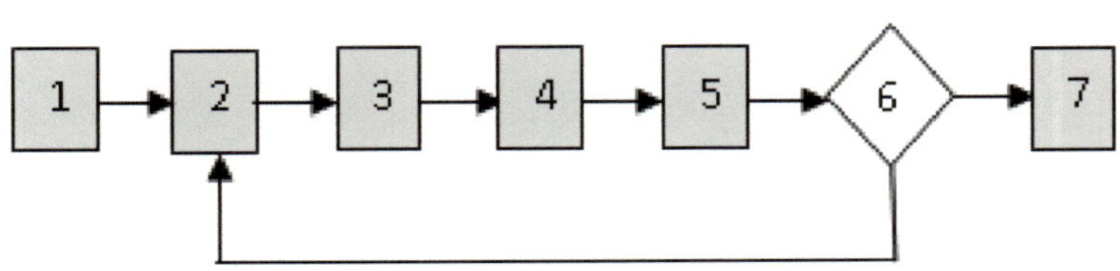

Figure 3.
Process loop of GA optimization method.

3. Mating the population to create progeny.

4. Mutate progeny.

5. Inserting new generated individuals into populations.

6. Are the system fitness function satisfied?

7. End search process for solution.

In this study, the aim of using GA optimization method is to tune the elements of the state weighting matrix Q and input weighting matrix R of the optimal LQR controller based on a selected fitness function which, should be minimised to a smallest value. The fitness function should be formulated based on the required performance characteristics. These optimized LQR elements are then employed to calculate the optimum values for PID controller gain parameters, which are used to stabilize the control system. The implementation procedure of the GA tuning method begins with the definition step of the chromosome representation. Each chromosome is represented by a strip of cells. Each cell corresponds to an element of the controller gain parameters. These cells are formed by real positive numbers and characterize the individual to be evaluated [13].

4. Hybrid PID control approaches

PID controller is a simple manipulating technique that can be successfully implemented for one dimension control systems. For multi dimensions systems it can use a multi channel PID controller system to control the dynamic behavior of these systems. Currently, there is a considerable interest by many researchers in development new control approaches using PID controller. Xiong and Fan [15] proposed a new adaptive PID controller based on model reference adaptive control (MRAC) concept for control of the DC electromotor drive. They presented an autotuning algorithm that combines PID control scheme and MRAC based on MIT rule to tune the controller parameters. Modified PI and PID controllers are introduced to regulate output voltage of DC-DC converters using MRAC manipulating technique [16, 17]. The parameters of the controllers are adapted effectively using MIT rule. Based on the adapted controllers parameters an improvement in the regulation behavior of the converters has been investigated.

Further improvement in the performance of the standard PID controller is also achieved by involving an integrator of order λ and differentiator of order μ to the controller structure based on Fractional Calculus and it is known as fractional order (FO) PID controller [7]. This extension could provide more flexibility in PID controller design and makes the system more robust, thus, enhancing its dynamic performance compared to its integer counterpart. In FOPID controller the manipulating parameters become five that provides more flexibility in the controller design and robust in the performance.

In the last decades, a new hybrid controller scheme using PID technology is proposed in [18–20] for different applications. The structure of the presented hybrid controller system is constructed by combination between conventional PID controller and state feedback LQR optimal controller. The gain parameters of the PID controller used to achieve desired output response are determined based on optimal LQR theory.

In this chapter, a hybrid PID controller based on LQR optimal technique is designed to stabilize 3DOF helicopter system. The proposed hybrid LQR-PID controller is optimized using GA optimization method, which is used to tune its gain parameters.

5. Case study: helicopter control system

5.1 Helicopter structure and modeling

The conceptual platform of 3DOF helicopter scheme is presented in **Figure 4**. It consists of an arm mounted on a base. The main body of the helicopter constructed of propellers driven by two motors mounted are the either ends of a short balance bar. The whole helicopter body is fixed on one end of the arm and a balance block installed at the other end.

The balance arm can rotate about the travel axis as well as slope on an elevation axis. The body of the helicopter is free to roll about the pitch axis. The system is provided by encoders mounted on these axes used to measure the travel motion of the arm and its elevation and pitch angle. The propellers with motors can generate an elevation mechanical force proportional to the voltage power supplied to the motors. This force can cause the helicopter body to lift off the ground. It is worth considering that the purpose of using a balance block is to reduce the voltage power supplied to the propellers motors. In this study, the nonlinear dynamics of 3DOF helicopter system is modelled mathematically based on developing the model of the system behavior for each of the axes.

5.1.1 Elevation axis model

The free body diagram of 3DOF helicopter system based on elevation axis is shown in **Figure 5**. The movement of the elevation axis is governed by the following differential equations:

$$J_\epsilon \ddot{\epsilon} = \left(J_p - J_\tau\right)\dot{\tau}^2 \cos\left(\epsilon\right) \sin\left(\epsilon\right) + l_1 F_m \cos\left(\rho\right) - l_1 F_m - M_G - M_{f,\epsilon} \qquad (17)$$

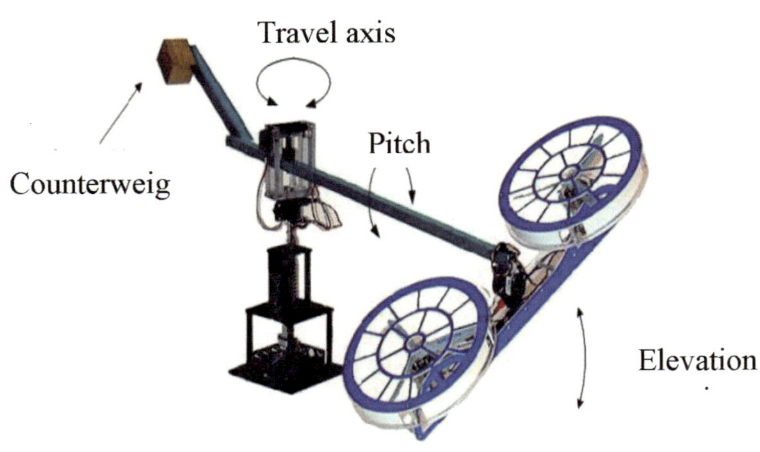

Figure 4.
Prototype model of 3DOF helicopter system.

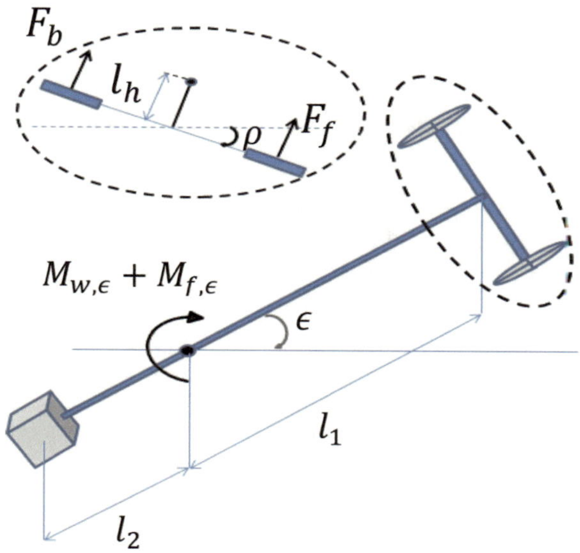

Figure 5.
Schematic diagram of elevation axis model for 3DOF helicopter system.

Where F_m is the thrust force of propeller motor and $M_{f,\epsilon}$ represents the torque component generated from combining the joint friction and air resistance. But the rotation angle of the pitch axis $\rho = 0$, if the elevation angle $\epsilon = 0$, then the torque exerted on the elevation axis will be zero. Eq. (17) based on Euler's second law becomes:

$$J_\epsilon \ddot{\epsilon} = l_1 F_m - M_{w,\epsilon} + M_{f,\epsilon} \tag{18}$$

$$J_\epsilon \ddot{\epsilon} = l_1 (F_f + F_b) - M_{w,\epsilon} + M_{f,\epsilon} \tag{19}$$

$$F_i = K_c V_i \; i = f, b \tag{20}$$

$$J_\epsilon \ddot{\epsilon} = K_c l_1 (V_f + V_b) - M_{w,\epsilon} + M_{f,\epsilon} \tag{21}$$

$$J_\epsilon \ddot{\epsilon} = K_c l_1 V_s - M_{w,\epsilon} + M_{f,\epsilon} \tag{22}$$

5.1.2 Pitch axis model

Consider the pitch schematic diagram of the system in **Figure 6**. It can be seen from the figure that the main torque acting on the system pitch axis is produced from the thrust force generated by the propeller motors. When $\rho \neq 0$, the gravitational force will also generate a torque $M_{w,\rho}$ acts on the helicopter pitch axis. The dynamics of the pitch axis can be modeled mathematically as follows:

$$J_\rho \ddot{\rho} = F_f l_\rho - F_b l_\rho - M_{w,\rho} - M_{f,\rho} \tag{23}$$

Where $M_{f,\rho}$ is the friction moment exerted on the pitch axis.

$$M_{w,\rho} = m_h g l_h \, sin\,(\rho) \, cos\,(\epsilon) \tag{24}$$

Based on the assumption that the pitch angle $\rho = 0$, $M_{w,\rho} = 0$, then Eq. (23) becomes as follows:

$$J_\rho \ddot{\rho} = l_\rho (F_f - F_b) - M_{f,\rho} \tag{25}$$

$$J_\rho \ddot{\rho} = K_c l_\rho (V_f - V_b) - M_{f,\rho} \tag{26}$$

$$J_\rho \ddot{\rho} = K_c l_\rho V_d - M_{f,\rho} \tag{27}$$

5.1.3 Travel axis model

The free body diagram of the helicopter system dynamics based on the travel axis is presented in **Figure 7**. In this model, when $\rho \neq 0$, the main forces acting on the helicopter dynamics are the thrust forces of propeller motors (F_f, F_b). These forces have a component that generates a torque on the travel axis. Assume that the helicopter body has roll up by an angle ρ as shown in **Figure 7**. Then the dynamics of travel axis for 3DOF helicopter system is modeled as follows:

$$J_r \dot{r} = -(F_f + F_b) \sin(\rho) l_1 - M_{f,r} \tag{28}$$

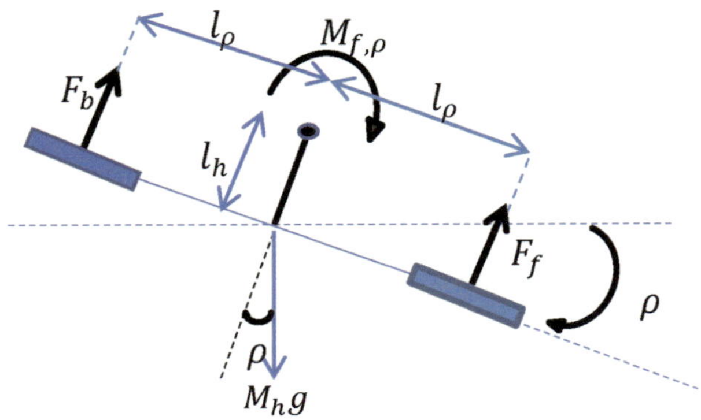

Figure 6.
Schematic diagram of the pitch axis model for 3DOF helicopter scheme.

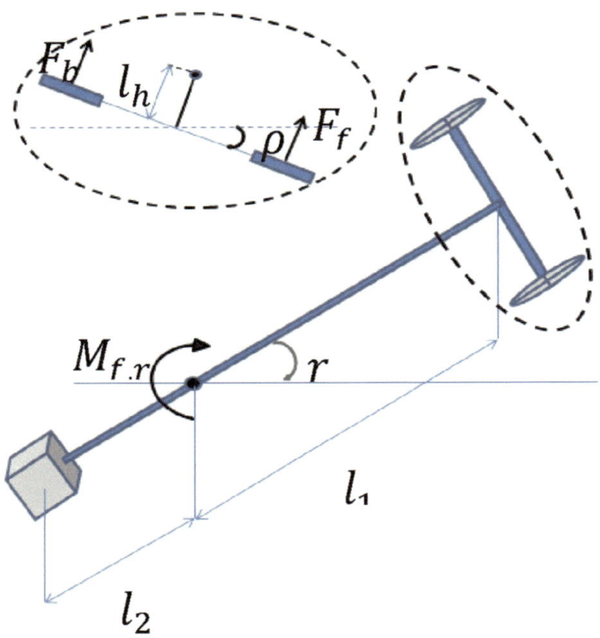

Figure 7.
Schematic diagram of the travel rate axis model for 3DOF helicopter scheme.

The thrust forces of the two propeller motors $(F_f + F_b)$ are required to keep the helicopter in flight case and is approximately W.

$$J_r\dot{r} = -Wsin(\rho)l_1 - M_{f,r} \tag{29}$$

Where $M_{f,r}$ is the friction moment exerted on the travel axis. As ρ approaches to zero, based on sinc function, $sin\ (\rho) = \rho$, the above equation becomes as follows:

$$J_r\dot{r} = -W\rho l_1 - M_{f,r} \tag{30}$$

Based on the assumption that the coupling dynamics, gravitational torque $(M_{w,\epsilon})$ and friction moment exerted on elevation, pitch and travel axis are neglected, then the dynamics modeling equations Eqs. (22), (27) and (30) for 3DOF helicopter system can be simplified as in Eqs. (31), (32) and (33) respectively [21].

$$\ddot{\epsilon} = \frac{K_c l_1}{J_\epsilon} V_s \tag{31}$$

$$\ddot{\rho} = \frac{K_c l_\rho}{J_\rho} V_d \tag{32}$$

$$\dot{r} = \frac{W l_1}{J_r} \rho \tag{33}$$

5.2 Helicopter state space model

In order to design a state feedback controller based on LQR technique for 3DOF helicopter system, the dynamics model of the system should be formulated in state space form. In this study, the proposed hybrid control algorithm is investigated for the purpose of control of pitch angle, elevation angle and travel rate of 3DOF helicopter scheme by regulating the voltage supplies to the front and back motors. Let $x(nx1) = [x_1, x_2, x_3, x_4, x_5, x_6, x_7]^T = [\epsilon, \rho, \dot{\epsilon}, \dot{\rho}, r, \mathfrak{z}, \gamma]^T$ be the state vector of the system, the state variables are chosen as the angles and rate and their corresponding angular velocities, and $\dot{\mathfrak{z}} = \epsilon, \dot{\gamma} = r$. The voltages supplied to the front and back propellers motors are considered the input's vector such that, $u(t)(\mathrm{mx1}) = [u_1, u_2]^{\mathrm{T}} = [V_f, V_b]^{\mathrm{T}}$ and the elevation angle, pitch angle and travel rate are assumed the output's vector such that, $y(t)(px1) = [\epsilon, \rho, r]^T$.

Based on Eqs. (31)-(33), choosing these state variables yields the following system state space model:

$$\begin{aligned}
\dot{x_1} &= \rho = x_2 \\
\dot{x_2} &= \dot{\epsilon} = x_3 \\
\dot{x_3} &= \ddot{\epsilon} = \frac{K_c l_1}{J_\epsilon}(V_f + V_b) \\
\dot{x_4} &= \ddot{\rho} = \frac{K_c l_\rho}{J_\rho}(V_f - V_b) \\
\dot{x_5} &= \dot{r} = \frac{W l_1}{J_r} x_2 \\
\dot{x_6} &= \dot{\mathfrak{z}} = x_1 \\
\dot{x_7} &= \dot{\gamma} = x_4
\end{aligned} \tag{34}$$

The general state and output matrix equations describing the dynamic behavior of the linear-time-invariant helicopter system in state space form are as follows:

$$x(t) = Ax(\dot{t}) + Bu(t) \tag{35}$$

$$y(t) = Cx(t) + Du(t) \tag{36}$$

Where $A(nxn)$ is the system matrix, $B(nxm)$ is the input matrix, $C(pxm)$ is the output matrix, and $D(mxp)$ is feed forward matrix, for the designed system. Based on Eqs. (34)–(36) are rewritten as follows [21].

$$\begin{bmatrix} \dot{\epsilon} \\ \dot{\rho} \\ \ddot{\epsilon} \\ \ddot{\rho} \\ \dot{r} \\ \dot{\mathfrak{z}} \\ \dot{\gamma} \end{bmatrix} = \begin{bmatrix} 0 & 0 & 1 & 0 & 0 & 0 & 0 \\ 0 & 0 & 0 & 1 & 0 & 0 & 0 \\ 0 & 0 & 0 & 0 & 0 & 0 & 0 \\ 0 & 0 & 0 & 0 & 0 & 0 & 0 \\ 0 & \frac{Gl_1}{J_t} & 0 & 0 & 0 & 0 & 0 \\ 1 & 0 & 0 & 0 & 0 & 0 & 0 \\ 0 & 0 & 0 & 0 & 1 & 0 & 0 \end{bmatrix} \begin{bmatrix} \epsilon \\ \rho \\ \dot{\epsilon} \\ \dot{\rho} \\ r \\ \mathfrak{z} \\ \gamma \end{bmatrix} + \begin{bmatrix} 0 & 0 \\ 0 & 0 \\ \frac{K_c l_1}{J_e} & \frac{K_c l_1}{J_e} \\ \frac{K_c l_\rho}{J_p} & -\frac{K_c \rho}{J_\rho} \\ 0 & 0 \\ 0 & 0 \\ 0 & 0 \end{bmatrix} \begin{bmatrix} V_f \\ V_b \end{bmatrix} \tag{37}$$

In this study, for the purpose of control system design, the model of the system is formulated in state space form using the physical parameters values listed in **Table 1** [21]. Based on Eq. (37) and using the parameters values in **Table 1**, the state equation of the system is given by Eq. (39):

$$\begin{bmatrix} \epsilon \\ \rho \\ r \end{bmatrix} = \begin{bmatrix} 1 & 0 & 0 & 0 & 0 & 0 & 0 \\ 0 & 1 & 0 & 0 & 0 & 0 & 0 \\ 0 & 0 & 0 & 0 & 1 & 0 & 0 \end{bmatrix} \begin{bmatrix} \epsilon \\ \rho \\ \dot{\epsilon} \\ \dot{\rho} \\ r \\ \mathfrak{z} \\ \gamma \end{bmatrix} + \begin{bmatrix} 0 & 0 \\ 0 & 0 \\ 0 & 0 \end{bmatrix} \begin{bmatrix} V_f \\ V_b \end{bmatrix} \tag{38}$$

$$\begin{bmatrix} \dot{\epsilon} \\ \dot{\rho} \\ \ddot{\epsilon} \\ \ddot{\rho} \\ \dot{r} \\ \dot{\mathfrak{z}} \\ \dot{\gamma} \end{bmatrix} = \begin{bmatrix} 0 & 0 & 1 & 0 & 0 & 0 & 0 \\ 0 & 0 & 0 & 1 & 0 & 0 & 0 \\ 0 & 0 & 0 & 0 & 0 & 0 & 0 \\ 0 & 0 & 0 & 0 & 0 & 0 & 0 \\ 0 & 2.065 & 0 & 0 & 0 & 0 & 0 \\ 1 & 0 & 0 & 0 & 0 & 0 & 0 \\ 0 & 0 & 0 & 0 & 1 & 0 & 0 \end{bmatrix} \begin{bmatrix} \epsilon \\ \rho \\ \dot{\epsilon} \\ \dot{\rho} \\ r \\ \mathfrak{z} \\ \gamma \end{bmatrix} + \begin{bmatrix} 0 & 0 \\ 0 & 0 \\ 5.8197 & 5.8197 \\ 63.949 & -63.949 \\ 0 & 0 \\ 0 & 0 \\ 0 & 0 \end{bmatrix} \begin{bmatrix} V_f \\ V_b \end{bmatrix} \tag{39}$$

5.3 Helicopter control system design

Based on step input, a hybrid controller is designed for the following desired performance parameters: rise time (t_r) less than 10 ms, settling time (t_s) less than 30 ms, maximum overshoot percentage, (M_O) less than 5%.

Under the assumption that the desired system states are zero the block diagram of the proposed helicopter control system based on the LQR controller is shown in

Symbol	Physical unit	Numerical values
J_ϵ	kg m^2	1.8145
J_r	kg m^2	1.8145
J_ρ	kg m^2	0.0319
W	N	4.2591
l_m	m	0.88
l_b	m	0.35
l_ρ	m	0.17
K_c	N/V	12

Table 1.
Values of physical parameters of 3DOF helicopter system.

Figure 8. The control system is analysed mathematically and then simulated using Matlab software tool to validate the proposed hybrid controller. Based on the desired performance parameters, which include rise and settling time, overshoot and error steady state, the fitness function of the control problem is formulated as follows:

$$F = 0.3S.t_r + 0.3S.t_s + 0.2S.O + 0.2S.e_{ss} \tag{40}$$

where, S is closed loop transfer function of the helicopter system, $S.t_r, t_s, O, e_{ss}$ are the rise time, settling time, maximum overshoot and error steady state of the closed-loop control system. It is worth considering that the control input effort is considered in the evaluation process of the proposed stabilizing helicopter system. In this study, the design of the controller is effectively optimized by using GA tuning method which is adopted to obtain optimum elements values for LQR weighting matrices Q and R. These optimized matrices are used to calculate the optimum controller gain matrix by using Eqs. (13) and (14). However, in this study, the LQR gain matrix is determined by using the Matlab command *'lqr'*.

5.3.1 PID approximation

In this subsection, the gain parameters K_p, K_i, K_d of the PID controller are calculated approximately from the feedback gain K of LQR controller based on GA tuning method. For this application, analyzing Eq. (15) yields the following control effort [22]:

$$\begin{bmatrix} u_1 \\ u_2 \end{bmatrix} = -\begin{bmatrix} k_{11} & k_{12} & k_{13} & k_{14} & k_{15} & k_{16} & k_{17} \\ k_{11} & -k_{12} & k_{13} & -k_{14} & -k_{15} & k_{16} & -k_{17} \end{bmatrix} x^T \tag{41}$$

where $x^T = \left[\epsilon, \rho, \dot{\epsilon}, \dot{\rho}, r, \mathfrak{z}, \gamma\right]^T$.

If ϵ_d, ρ_d and r_d are the desired pitch angle, elevation angle and travel rate of the helicopter system, it can express the form of PID controllers used to meet the desired output states as follows: [7, 8].

In this study, for elevation angle, the control equation is based on the following PID control equation:

$$V_s = K_{\epsilon p} e_\epsilon + K_{\epsilon d} \dot{e_\epsilon} + K_{\epsilon i} \int e_\epsilon dt \tag{42}$$

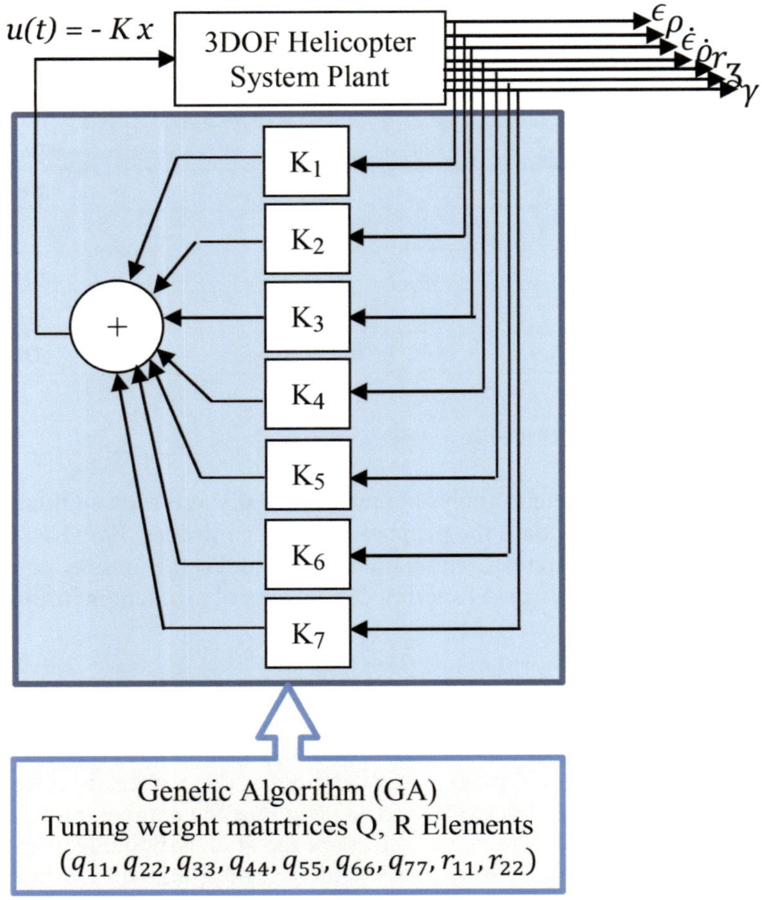

Figure 8.
LQR controller based on GA for 3DOF helicopter system.

$$V_s = K_{\epsilon p}(\epsilon_d - \epsilon) - K_{\epsilon d}\dot{\epsilon} + K_{\epsilon i}\int(\epsilon_d - \epsilon)dt \tag{43}$$

While the pitch angle is controlled by the following PD control equation:

$$V_d = K_{\rho p}e_\rho + K_{\rho d}\dot{e_\rho} \tag{44}$$

$$V_d = K_{\rho p}(\rho_d - \rho) - K_{\rho d}\dot{\rho} \tag{45}$$

The travel rate is gonverned by the following PI control equation:

$$\rho_d = K_{rp}e_r + K_{ri}\int e_r dt \tag{46}$$

$$\rho_d = K_{rp}(r_d - r) + K_{ri}\int(r_d - r)dt \tag{47}$$

Where $e_\epsilon = \epsilon_d - \epsilon, e_\rho = \rho_d - \rho, e_r = r_d - r, \dot{e_\epsilon} = -\dot{\epsilon}$ and$\dot{e_\rho} = -\dot{\rho}$.

5.3.2 Elevation control using PID controller

Summing the rows of (41) results the following [21]:

$$u_1 + u_2 = -(2k_{11}\epsilon + 2k_{13}\dot{\epsilon} + 2k_{16}\xi) = -\left(2k_{11}\epsilon + 2k_{13}\dot{\epsilon} + 2k_{16}\int \epsilon dt\right) \tag{48}$$

The above equation can be written as

$$V_s = 2k_{11}(\epsilon_d - \epsilon) - 2k_{13}\dot{\epsilon} + 2k_{16}\int(\epsilon_d - \epsilon)dt \tag{49}$$

It is obvious that Eqs. (43) and (49) have the same structure, this means that the gain parameters of the pitch PID controller can be obtained from the gain elements of the LQR controller. Thus, comparing Eq. (43) with Eq. (49), yields the following gain relationships:

$$\begin{aligned} K_{\epsilon p} &= 2k_{11} \\ K_{\epsilon d} &= 2k_{13} \\ K_{\epsilon i} &= 2k_{16} \end{aligned} \tag{50}$$

The block diagram of closed-loop control system for 3DOF helicopter system based on hybrid LQR-PID controller is shown in **Figure 9**. Taking Laplace transform for elevation axis model Eq. (31) yields the following equation:

$$J_e\epsilon(s).s^2 = K_c l_1 V_s(s) \tag{51}$$

The transfer function of the elevation axis plant is given by:

$$\frac{\epsilon(s)}{V_s(s)} = \frac{K_c l_1}{J_e s^2} \tag{52}$$

The transfer function of the PID controller is as follows:

$$\frac{V_s(s)}{E_\epsilon(s)} = \frac{K_{\epsilon d}s^2 + K_{\epsilon p}s + K_{\epsilon i}}{s} \tag{53}$$

where $E_\epsilon(s) = \epsilon_d(s) - \epsilon(s)$, the open loop transfer function of the elevation axis control $G_\epsilon(s)$ is given by:

$$G_\epsilon(s) = \frac{\epsilon(s)}{E_\epsilon(s)} = \frac{V_s(s)}{E_\epsilon(s)}\frac{\epsilon(s)}{V_s(s)} \tag{54}$$

Based on Eqs. (52) and (53), the open loop elevation transfer function becomes:

$$G_\epsilon(s) = \frac{K_c l_1}{J_e s^2}\frac{K_{\epsilon d}s^2 + K_{\epsilon p}s + K_{\epsilon i}}{s} \tag{55}$$

The closed loop transfer function for elevation angle control is as follows:

$$\frac{\epsilon(s)}{\epsilon_c(s)} = \frac{G_\epsilon(s)}{1 + G_\epsilon(s)} = \frac{K_c l_1 K_{\epsilon d}s^2 + K_c l_1 K_{\epsilon p}s + K_c l_1 K_{\epsilon i}}{J_e s^3 + K_{\epsilon d}s^2 + K_{\epsilon p}s + K_{\epsilon i}} \tag{56}$$

5.3.3 Pitch control using PD controller

Similarly, the difference of the rows of Eq. (41) results in

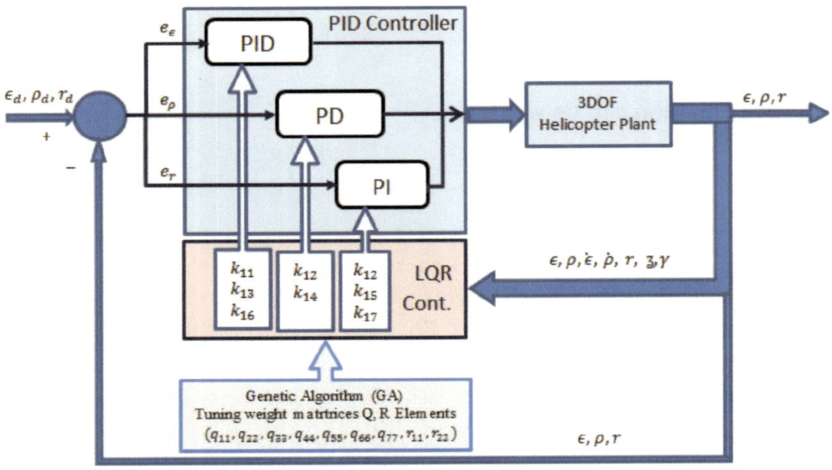

Figure 9.
Control system block diagram for helicopter elevation, pitch and travel axis using PID controller.

$$u_1 - u_2 = -2k_{12}\rho - 2k_{14}\dot{\rho} - 2k_{15}(r - r_d) - 2k_{17}\gamma \tag{57}$$

$$V_d = -2k_{12}\rho - 2k_{14}\dot{\rho} - 2k_{15}(r - r_d) - 2k_{17}\int(r - r_d)dt \tag{58}$$

Substitution Eq. (47) in Eq. (45) results,

$$V_d = -K_{\rho p}\rho - K_{\rho d}\dot{\rho} + K_{\rho p}K_{rp}(r_d - r) + K_{\rho p}K_{ri}\int(r_d - r)dt \tag{59}$$

It is clear that Eqs. (58) and (59) have exactly the same structure. Then, by comparing these equations, it can obtain the feedback gains for the PID controller from the LQR gains parameters as follows:

$$\begin{aligned} K_{\rho p} &= 2k_{12} \\ K_{\rho d} &= 2k_{14} \\ K_{rp} &= \frac{k_{15}}{k_{12}} \\ K_{ri} &= \frac{k_{17}}{k_{12}} \end{aligned} \tag{60}$$

Taking Laplace transform for pitch axis model Eq. (32) yields:

$$J_\rho \rho(s)s^2 = K_c l_\rho V_d(s) \tag{61}$$

The transfer function for pitch axis model is given by:

$$\frac{\rho(s)}{V_d(s)} = \frac{K_c l_\rho}{J_\rho s^2} \tag{62}$$

The transfer function of the PD controller is as follows:

$$\frac{V_d(s)}{E_\rho(s)} = K_{\rho d}s + K_{\rho p} \tag{63}$$

where $E_\rho(s) = \rho_d(s) - \rho(s)$, based on Eqs. (62) and (63) the open loop transfer function of the pitch axis control is given by:

$$G_p(s) = \frac{\rho(s)}{E_\rho(s)} = \frac{\rho(s)}{V_d(s)}\frac{V_d(s)}{E_\rho(s)} = \frac{K_c l_\rho (K_{\rho d}s + K_{\rho p})}{J_e s^2} \tag{64}$$

The closed loop transfer function of pitch angle is given by:

$$\frac{\rho(s)}{\rho_c(s)} = -\frac{K_c l_\rho K_{\rho d}s + K_c l_\rho K_{\rho p}}{J_\rho s^2 + K_c l_\rho K_{\rho d}s + l_\rho K_{\rho p}} \tag{65}$$

5.3.4 Travel control using PI controller

Taking Laplace transform for travel axis model Eq. (33) results:

$$r(s)s = \frac{Wl_1}{J_r}\rho_d(s) \tag{66}$$

The transfer function for travel axis model is given by:

$$\frac{r(s)}{\rho_d(s)} = \frac{Wl_1}{J_r\, s} \tag{67}$$

The transfer function of the PI controller is as follows:

$$\frac{\rho_\rho(s)}{E_r(s)} = K_{rp} + \frac{K_{ri}}{s} \tag{68}$$

where $E(s) = \rho_d(s) - \rho(s)$, the open loop transfer function of the travel axis control is given by:

$$G_r(s) = \frac{r(s)}{\rho_d(s)}\frac{\rho_\rho(s)}{E_r(s)} = \frac{Wl_1(K_{rp}s + K_{ri})}{J_r s^2} \tag{69}$$

The closed loop transfer function for travel angle is as follows:

$$\frac{r(s)}{r_d(s)} = \frac{Gl_1K_{rp}s + Gl_1K_{ri}}{J_t s^2 + Gl_1K_{rp}s + Gl_1K_{ri}} \tag{70}$$

5.4 Controller simulation and results

5.4.1 GA-LQR controller

In order to validate the proposed helicopter stabilizing system, the LQR controller is analysed mathematically using Matlab tool. Based on objective function *(J)* and using the Matlab command *"lqr"* the elements of the LQR weighting matrices *Q*, *R* are tuned using GA optimization method. For this application, each chromosome in GA tuning approach is represented by nine cells which correspond to the weight matrices elements of the LQR controller as shown in **Figure 10**. By this representation it can adjust the LQR elements in order to achieve the required performance. The parameters of the GA optimization approach chosen for the tuning process of the helicopter control system are listed in **Table 2**. Converging

q_{11}	q_{22}	q_{33}	q_{44}	q_{55}	q_{66}	q_{77}	r_{11}	r_{22}

Figure 10.
Definition of GA chromosome.

elements of the LQR weight matrices Q and R through iteration based on GA optimization method are presented in **Figure 11**.

Based on the proposed fitness function stated in Eq. (40), the LQR weighting matrices Q and R obtained based on the GA tuning approach are given by:

$$\begin{aligned} Q &= blkdig(26.258, 0.869, 0.431, 0.475, 1.87, 0.026, 0.705), \\ R &= blkdig(0.469, 0.469), \end{aligned}$$

The feedback gain matrix of the LQR controller can be mathematically calculated using Eq. (13), where P matrix is the stabilizing solution of the Riccati equation stated in Eq. (14).

In this application, by using the state matrix (A), input matrix (B) and the tuned weighting matrices (Q, R), the optimized feedback gain matrix K stated below is determined using the Matlab software instruction:

$$K = lqr(A, B, Q, R)$$

$$K = \begin{bmatrix} 5.3232 & 2.6817 & 1.1719 & 0.7399 & 2.0590 & 0.1651 & 0.8661 \\ 5.3232 & -2.6817 & 1.1719 & -0.7399 & -2.0590 & 0.1651 & -0.8661 \end{bmatrix}$$

Based on the feedback gain matrix and using Eq. (11), the LQR control effort vector for the 3DOF helicopter system is dertermined as follows:

$$\begin{bmatrix} u_1 \\ u_2 \end{bmatrix} = -\begin{bmatrix} 5.323 & 2.682 & 1.172 & 0.739 & 2.059 & 0.165 & 0.866 \\ 5.323 & -2.682 & 1.172 & -0.739 & -2.059 & 0.165 & -0.866 \end{bmatrix} x^T \tag{71}$$

5.4.2 GA-PID controller

Based on Eqs. (50), (60) and (71), the absolute values of PID, PD and PI gain parameters for elevation, pitch and travel axis model respectively for helicopter system are listed in **Table 3** [21]. Using the values in **Tables 1** and **3**, the closed-loop transfer function of elevation, pitch and travel axis Eqs. (56), (65) and (70) become as in Eqs. (72), (73) and (74) respectively:

$$\frac{\epsilon(s)}{\epsilon_c(s)} = \frac{1445s^3 + 247\,s + 674.9}{1.8145s^3 + 1445s^2 + 2474s + 674.9} \tag{72}$$

GA property	Value/Method
Population Size	20
Max No. of Gen.	100
Selection Method	Normalized Geo. Selection
Crossover Method	Scattering
Mutation Method	Uniform Mutation

Table 2.
Parameters of GA tuning method.

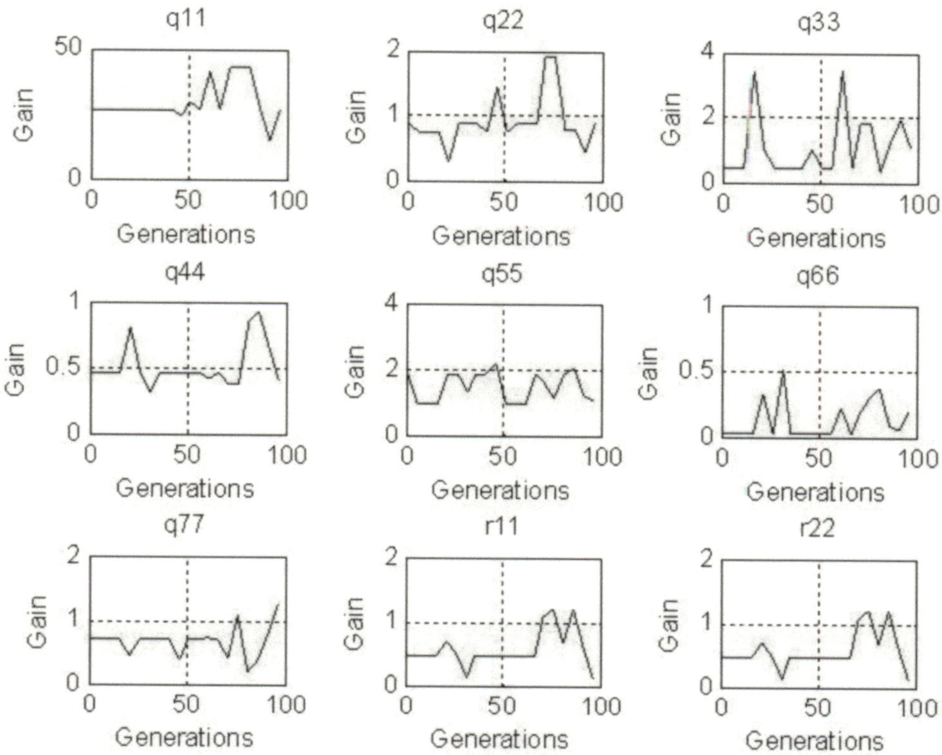

Figure 11.
Number of generation of GA-LQR parameters Q and R.

PID parameters	Relationship	Absolute Value
K_{ep}	$2k_{11}$	10.6463
K_{ed}	$2k_{13}$	2.3438
K_{ei}	$2k_{16}$	0.3302
K_{pp}	$2k_{12}$	5.3634
K_{pd}	k_{14}	1.4799
K_{rp}	$\frac{k_{15}}{k_{12}}$	0.7678
K_{ri}	$\frac{k_{17}}{k_{12}}$	0.3230

Table 3.
Values of gain parameters for PID, PD and PI controllers.

$$\frac{p(s)}{p_c(s)} = \frac{69.08\,s + 269.3}{0.0319s^2 + 69.08\,s + 269.3} \tag{73}$$

$$\frac{r(s)}{r_c(s)} = \frac{2.878\,s + 1.634}{1.815\,s^2 + 2.878\,s + 1.634} \tag{74}$$

Based on bounded input signal, the elevation, pitch and travel axis model of 3DOF helicopter system are unstable as they give unbounded outputs. The output responses for elevation, pitch and travel angle are illustrated in the **Figure 12**. It can be say that the open loop helicopter system without control action is unable to provide a stable output response.

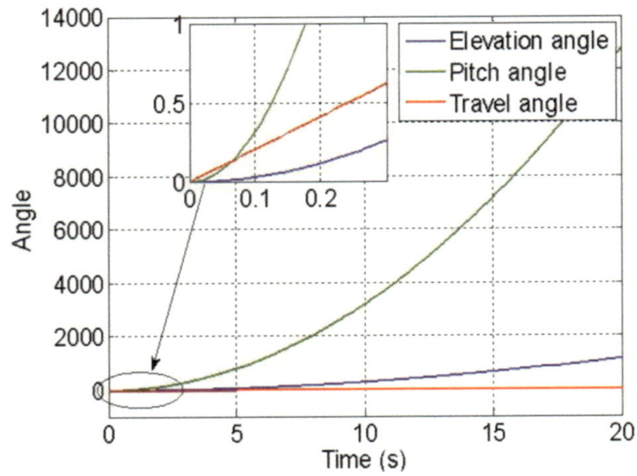

Figure 12.
Open loop response of Helicopter system.

In this study, in order to achieve a stable output, a hybrid control system using LQR based PID controller for 3DOF helicopter system is proposed to control the dynamic behaviour of the system. To validate the proposed helicopter stabilization system, the controller is simulated using Matlab programming tool. Three axis, elevation, pitch, travel rate, are considered in the simulation process of the control system. The performance of the helicopter balancing system is evaluated under unit step reference input using rise, settling time overshoot and steady state error parameters for the elevation, pitch and travel angles to simulate the desired command given by the pilot.

5.4.2.1 Elevation LQR-PID controller

This section deals with the simulation of LQR based PID controller used to control the position of helicopter elevation model. The parameters of the hybrid controller are tuned using GA optimization method. **Figure 13** presents a tracking control curve of the demand input based on the PID controller using optimized gain parameters listed in **Table 3** for helicopter elevation angle.

The simulation results show that the controller successed to guide the system output through the desired input trajectory effectively with negligible overshoot, short rise and settling time of 0.1 ms and 0.3 ms respectively.

5.4.2.2 Pitch LQR-PD controller

In this section, an optimized LQR-PD controller based on GA tuning approach is designed to control the dynamic model of helicopter pitch angle. Based on the optimized PD parameters stated in **Table 3**, the output response of the proposed helicopter tracking system is illustrated in **Figure 14**. It is obvious from the minifigure of the system response that the LQR-PD controller succeeded to force the pitch angle state of the helicopter system to follow the desired trajectory effectively without overshoot, shorter rise and settling time and zero steady state tracking error.

5.4.2.3 Travel LQR-PI controller

The control of the travel rate for the 3DOF helicopter system is governed by a GA-LQR based PI controller. The time response of the optimized PI tracking system

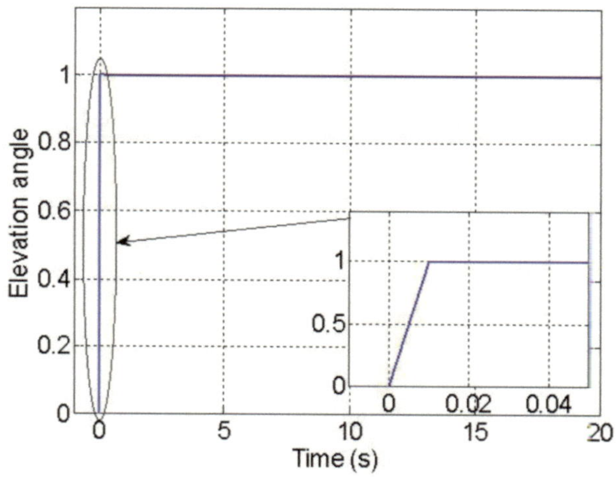

Figure 13.
Closed-loop response of the elevation model system.

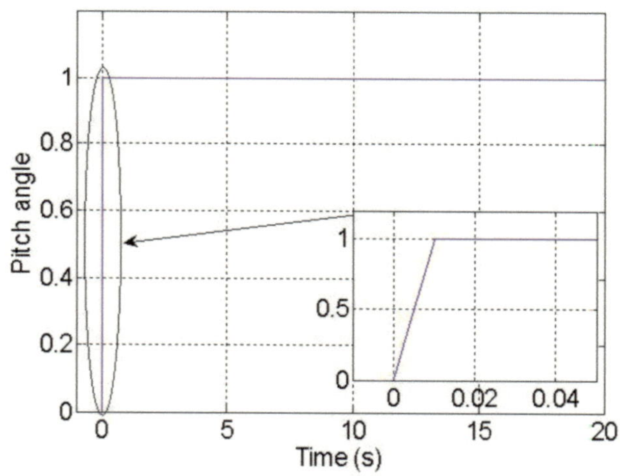

Figure 14.
Closed-loop response of the pitch model system.

using optimum gain parameters which are listed in **Table 3**. is shown in **Figure 15**. It can be noted from the miniplot of the system response that the optimised hybrid LQR-PI controller enabled the system output state to track the desired input trajectory without overshoot, and shorter rise and settling time with minimal steady state tracking error.

The control inputs supplied to the propellers motors of the proposed 3DOF helicopter system are shown in **Figure 16**. Consequently, it can say that the control performance of optimised GA-LQR based PID, PD and PI controllers for helicopter elevation, pitch and travel axis model respectively was acceptable through tracking the system output states for the reference input efficiently. Based on the minifigures of **Figures 13** and **14** and **Figure 15**, the performance parameters of PID, PD and PI controller for helicopter elevation, pitch and travel axis are listed in **Table 4**. From the response data of the controlled helicopter system in **Table 4** it can be said that the hybrid controllers were able to provide robust and good tracking performance in both the transient and steady state responses.

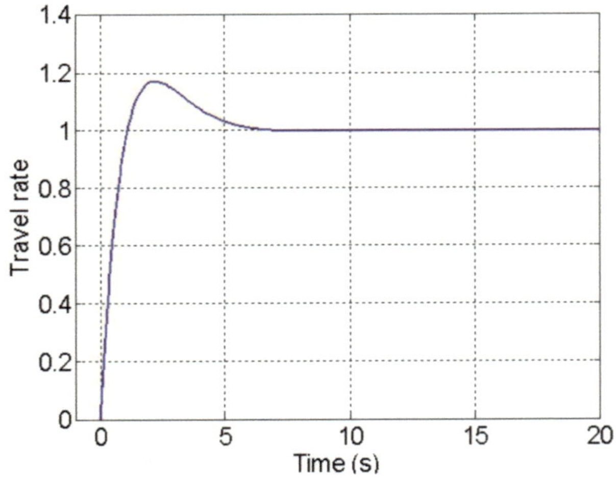

Figure 15.
Closed-loop response of the travel model system.

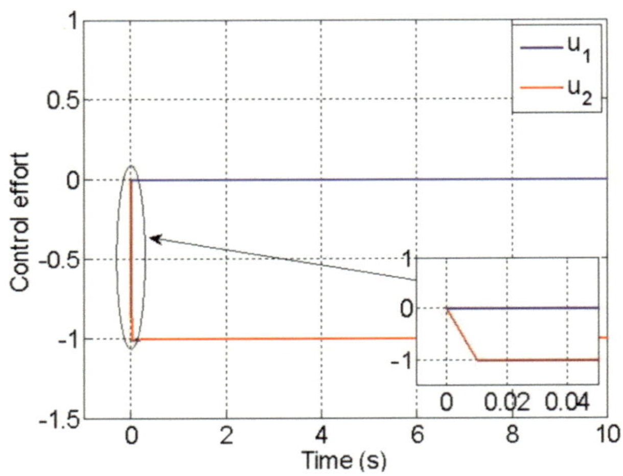

Figure 16.
Conrol input of 3DOF helicopter control system.

Controller	t_r(s)	t_s(s)	M_o %
Elevation PID	0.343	0.535	1.1
Pitch PD	0.582	1.05	0
Travel PI	1.17	12.4	5.29

Table 4.
Values of performance elements s of controllers.

6. Conclusions

In this study, a new hybrid control methodology has been developed for complex dynamical systems through combinig the LQR optimal technique with traditional PID controller. An efficient hybrid control system has been designed to stabilize 3DOF helicopter systems. The dynamics of elevation, pitch and travel axis for a helicopter system is modeled mathematically and then formulated in state space form to enable utilizing state feedback controller technique. In the proposed helicopter stabilizing scheme, a combination of a conventional PID control with LQR state feedback controller is adopted to stabilize the elevation, pitch and travel axis of the helicopter scheme. The gain parameters of the traditional PID controller are determined from the gain matrix of state feedback LQR controller. In this research, the LQR controller is optimized by using GA tuning technique. The GA optimization method has been adopted to find optimum values for LQR gain matrix elements which are utilized to find best PID gain parameters. The output response of the optimized helicopter control system has been evaluated based on rise time, setting time, overshoot and steady state error parameters. The simulation results have shown the effectiveness of the proposed GA-LQR based PID controller to stabilize the helicopter system at desired values of the elevation and pitch angle and travel parameters.

Author details

Ibrahim K. Mohammed
Systems and Control Engineering Department, College of Electronics Engineering, Ninevah University, Mosul, Iraq

*Address all correspondence to: ibrahim.mohammed@uoninevah@.edu.iq

References

[1] Lotfollahzade M, Akbarimajd A, Javidan J. Design LQR and PID Controller for Optimal Load Sharing of an Electrical Microgrid. Int. Res. J. Appl. Basic Sci. 2013;**4**:704-712 https://europub.co.uk/articles/5372

[2] Lindiya S, Subashini N, Vijayarekha K. Cross Regulation Reduced Optimal Multivariable Controller Design for Single Inductor DC-DC Converters. Energies. 2019;**12**: 477. DOI: 10.3390/en12030477

[3] Sen M A, Kalyoncu M. Grey Wolf Optimizer Based Tuning of a Hybrid LQR-PID Controller for Foot Trajectory Control of a Quadruped Robot. Gazi Univ. J. Sci. 2019, 32, 674–684. https://www.researchgate.net/publication/335015963_Grey_Wolf_Optimizer_Based_Tuning_of_a_Hybrid_LQR-PID_Controller_for_Foot_Trajectory_Control_of_a_Quadruped_Robot

[4] Nagarkar M, Bhalerao Y, Patil GV, Patil RZ. Multi-Objective Optimization of Nonlinear Quarter Car Suspension System PID and LQR Control. Procedia Manuf. 2018;**20**:420-427. DOI: 10.1016/j.promfg.2018.02.061

[5] Mohammed I, Sharif B, Neasham J. Design and implementation of a magnetic levitation control system for robotically actuated capsule endoscopes. In: Proceedings International Symposium on Robotic and Sensors Environments (ROSE2012); 16-18 November 2012; Magdeburg. Germany: IEEE; 2012. pp. 140-145

[6] Mohammed I, Sharif B, Neasham J. Design and implementation of positioning and control systems for capsule endoscopes. IET Journal of Science, Measurement and Technology. 2020;14(7): 745-754. https://digital-library.theiet.org/content/journals/10.1049/iet-smt.2019.0223

[7] Mohammed I, Abdulla A. Fractional Order PID Controller Design for Speed Control DC Motor based on Artificial Bee Colony Optimization. International Journal of Computer Applications. 2018; 179(24); 43-49. https://www.researchgate.net/publication/323873116_Fractional_Order_PID_Controller_Design_for_Speed_Control_DC_Motor_based_on_Artificial_Bee_Colony_Optimization

[8] Abdelhakim I, Madjid K, Yassine B, Sid Ahmed T. Speed Control of DC Motor Using PID and FOPID Controllers Based on Differential Evolution and PSO. International Journal of Intelligent engineering and Systems. 2018;**11**(4):241-249. DOI: 10.22266/ijies2018.0831.24

[9] Ahmad A. at. el. Control of magnetic levitation system using fuzzy logic control. In: Proceedings of the 2nd IEEE International Conference on Computational Intelligence; 28-30 September. Vol. 2010. Tuban. Indonesia: IEEE; 2010. pp. 51-56

[10] Jain R, Aware MV, Junghare. Tuning of Fractional Order PID Controller Using Particle Swarm Optimization Technique for DC Motor Speed Control. In: Proceedings International Conference on Power Electronics, Intelligent Control and Energy Systems; 4-6 July 2016; Delhi. India: IEEE; 2016; p. 1-4

[11] Mohammed I. Design and Simulation of Three Degrees-of-Freedom Tracking Systems for Capsule Endoscope. Journal of Dynamic Systems, Measurements, and Control. 2016;138: 111002-1-11102-11. DOI: 10.1115/1.4033830

[12] Mohammed I, Abdulla A. Design of optimised linear quadratic regulator for capsule endoscopes based on artificial bee colony tuning algorithm.

International Journal for Engineering Modeling. 2018;**31**(12):77-98 https://doi.org/10.31534/engmod.2018.1-2.ri.02_vjan

[13] Abdulla A, Mohammed I, Jasim A. Roll Control System Design Using Auto Tuning LQR Technique. International Journal of Engineering and Innovative Technology. 2017;**6**(2):11-22 https://www.ijeit.com/Vol%207/Issue%201/IJEIT1412201707_02.pdf

[14] Mohammed I, Abdulla A, Ahmed J. Speed Control of DC Motor Using MRAC and Genetic Algorithm based PID Controller. International Journal of Industrial Electronics and Electrical Engineering. 2020;**8**(1):1-6 http://www.ijieee.org.in/paper_detail.php?paper_id=16837&name=Speed_Control_of_DC_Motor_using_MRAC_and_Genetic_Algorithm_based_PID_Controller

[15] Xiong A, Fan Y. Application of a PID Controller using MRAC Techniques for Control of the DC Electromotor Drive. In: Proceedings IEEE International Conference on Mechatronics and Automation; 5-8 Aug 2007; Harbin China: IEEE. 2007. pp. 2616-2626

[16] Ardhenta L, Subroto RK. Application of direct MRAC in PI controller for DC-DC boost converter. International Journal of Power Electronics and Drive System (IJPEDS). 2020;**11**(2):551-858 http://ijpeds.iaescore.com/index.php/IJPEDS/article/view/20543

[17] Ardhenta L, Kuswiradyo P, Subroto RK. Model Direct Adaptive Control of Buck Converter Using MRAC. International Journal of Innovative Technology and Exploring Engineering (IJITEE). 2019;**8**(12):1108-1112 https://www.ijitee.org/wpcontent/uploads/papers/v8i12/L38821081219.pdf

[18] Kumar EV. Jerome J. LQR based optimal tuning of PID controller for trajectory tracking of Magnetic Levitation System. In: Proceedings: International Conference On DESIGN AND MANUFACTURING. Procedia Engineering; 64(2013). 2013. pp. 254-264

[19] Saleem O, Rizwan M. Performance optimization of LQR-based PID controller for DC-DC buck converter via iterative-learning-tuning of state-weighting matrix. International Journal of Numerical Modeling Electronic Networks, Drives and Fields. 2019;32(3); 1-17. https://onlinelibrary.wiley.com/doi/epdf/10.1002/jnm.2572

[20] Jian-Bo H, Qing-Guo W, Tong-Heng L. PI/PID controller tuning via LQR approach. In: Proceedings the 37th IEEE Conference on Decision and Control; 18-18 Dec 1998. Tampa USA: IEEE; 1998. pp. 1177-1182

[21] Choudhary S K. LQR Based PID Controller Design for 3-DOF Helicopter System. International Journal of Electrical and Information Engineering. 2014;8)8(: 1498-1503.

[22] Fan L, Joo E. Design for Auto-tuning PID Controller Based on Genetic Algorithms. In: Proceedings IEEE 4th International Conference on Industrial Electronics and Applications (ICIEA); 25-27 May 2009; Xi'an China: IEEE. 2009. pp. 1924-1928

Chapter 3

Improving Disturbance-Rejection by Using Disturbance Estimator

Damir Vrančić and Mikuláš Huba

Abstract

The main tasks of control in various industries are either tracking the setpoint changes or rejecting the process disturbances. While both aim at maintaining the process output at the desired setpoint, the controller parameters optimised for setpoint tracking are generally not suitable for optimal disturbance rejection. The overall control performance can be improved to some extent by using simpler 2-DOF PID controllers. Such a controller structure allows the disturbance rejection to be optimised, while it also improves the setpoint tracking performance with additional controller parameters (usually through the setpoint weighting factors). Since such 2-DOF structures are usually relatively simple, the optimization of tracking performance is usually limited to the reduction of process overshoots instead of achieving an optimal (fast) tracking response. In this chapter, an alternative approach is presented in which the parameters of the PID controller are optimised for reference tracking, while the performance of the disturbance rejection is substantially increased by introducing a simple disturbance estimator approach. The mentioned estimator requires adding two simple blocks to the PID controller. The blocks are the second-order transfer functions whose parameters, including the PID controller parameters, can be calculated analytically from the process characteristic areas (also called process moments). The advantage of such an approach is that the mentioned areas can be analytically calculated directly from the process transfer function (of any order with time delay) or from the time response of the process when the steady state of the process is changed. Both of the above calculations are absolutely equivalent. Moreover, the output noise of the controller is under control as it is considered in the design of the controller and compensator. The closed loop results on several process models show that the proposed method with disturbance estimator has excellent tracking and disturbance rejection performance. The proposed controller structure and tuning method also compare favourably with some existing methods based on non-parametric description of the process.

Keywords: tuning method, disturbance rejection, disturbance estimator, multi-objective design

1. Introduction

The control of industrial processes requires efficient control loops. A majority of the control loops in various industries are implemented by the Proportional-Integrative-Derivative (PID) control algorithms. For efficient control, the PID controllers require proper tuning of the PID controller parameters. The parameters can be calculated to optimise various performance criteria such as integral of error (IE), integral of absolute error (IAE), integral of squared error (ISE) and similar [1–4]. However, the most important decision that should be made in advance is the choice of the main purpose of the closed-loop system. Namely, the user should choose between the optimal closed-loop responses to reference changes (so-called tracking responses) or the optimal response to process disturbances. While there are many industrial processes that require optimal reference tracking responses, such as robot manipulation, welding, and batch processes, the majority of industrial processes require optimal disturbance rejection.

The history of tuning rules is long, originating in the 1940s with the famous Ziegler-Nichols tuning rules. In the following decades, many other tuning rules have been developed [1, 2, 4–10]. The rules can be generally categorised according to the required data of the process. The process can be described either in parametric form, e.g., as a process model (transfer function), or in nonparametric form, e.g., as a process time-response.

A relatively new tuning method that optimises either closed-loop tracking or disturbance rejection is the Magnitude-Optimum-Multiple-Integration (MOMI) method [7, 9, 11, 12]. The MOMI method is based on the Magnitude Optimum method, which aims to optimise the frequency response of the closed loop to achieve fast and stable closed loop time response [10, 13–15]. An interesting feature of the MOMI method is that it works either on the process given by its transfer function (of arbitrary order with time delay) or directly on the time response of the process during the steady state change. It is worth noting that both the parametric and non-parametric process data give exactly the same PID tuning results.

Many tuning methods for PID controllers provide different sets of controller parameters for tracking and disturbance rejection response. Similarly, the MOMI method primarily optimises the tracking response, while its modification, the Disturbance-Rejection-Magnitude-Optimum (DRMO) method, aims at optimising the disturbance rejection response. The latter significantly improves the disturbance rejection response, while the tracking response slows down due to the implemented reference-weighting gain or reference signal filter [9, 16, 17].

The main approach presented in this chapter is the alternative approach. First, the parameters of the PID controller are optimised for tracking performance. Then, a simple disturbance estimator is introduced to significantly increase the disturbance rejection performance [18, 19]. The advantages of the above approach are twofold. First, the disturbance rejection performance can significantly outperform that obtained by the DRMO method. Second, the parameters of the disturbance estimator can also be obtained directly from the non-parametric process data in the time domain. Therefore, the proposed approach can still be applied to the process data which is either in parametric or non-parametric form.

However, in practice, the process output noise is always present. If the controller or estimator gains are too high, the process input signals may be too noisy for practical applications. Therefore, noise attenuation should already be taken into account when calculating the controller and estimator parameters. This chapter shows how to achieve the best trade-off between performance and noise attenuation.

2. Process and controller description

The classic 1-degree-of-freedom (1-DOF) control loop configuration of the process and the controller is shown in **Figure 1**, where the signals r, e, u, d and y

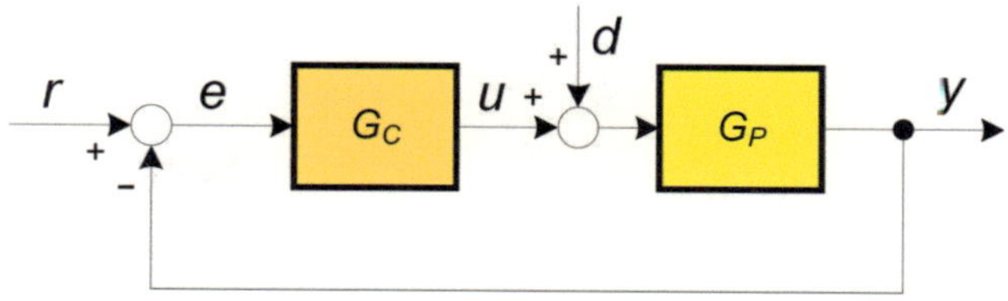

Figure 1.
The 1-DOF PID controller and the process in the closed-loop configuration.

represent the reference, the control error, the controller output, the process input disturbance, and the process output, respectively.

A process model (1) can be described by the following process transfer function:

$$G_P(s) = \frac{K_{PR}(1 + b_1 s + b_2 s^2 + \dots + b_m s^m)}{1 + a_1 s + a_2 s^2 + \dots + a_n s^n} e^{-sT_{del}} \quad (1)$$

where a_1 to a_n are the denominator coefficients, b_1 to b_m are the numerator coefficients, K_{PR} is the process gain, and T_{del} is the process time delay. Note that $n > m$ represents a strictly proper process transfer function and that the process is stable.

The PID controller is described by the following expression:

$$G_C(s) = \frac{K_I + K_P s + K_D s^2}{s(1 + sT_F)} \quad (2)$$

where K_I is the integrating gain, K_P is the proportional gain, and K_D is the derivative gain. Note that all three controller terms are filtered by the first-order filter with time constant T_F.

The closed-loop transfer function G_{CL} between the reference (r) and the process output (y) is as follows:

$$G_{CL} = \frac{G_C G_P}{1 + G_C G_P} \quad (3)$$

Since the structure of a 1-DOF PID controller does not provide optimal tracking and disturbance rejection at the same time, the 2-degrees-of-freedom (2-DOF) controller can be used instead [1, 2, 4, 8, 16, 20], where G_{CR} and G_{CY} denote the controller transfer function from the reference and the process output, respectively:

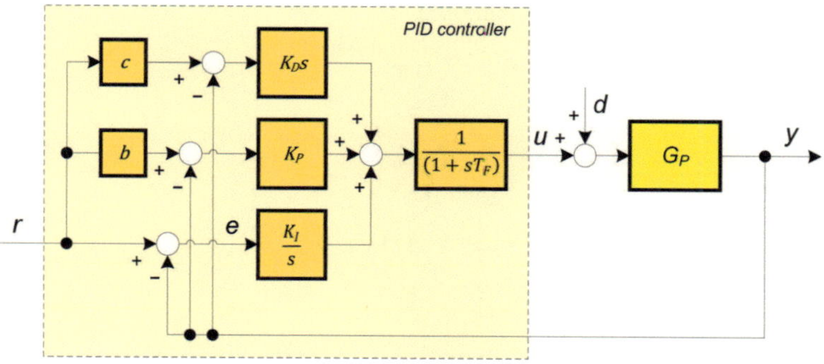

Figure 2.
The 2-DOF PID controller and process in the closed-loop configuration.

$$
\begin{aligned}
u &= G_{CR}(s)r - G_{CY}(s)y \\
G_{CR} &= \frac{K_I + bK_P s + cK_D s^2}{s(1+sT_F)} \\
G_{CY} &= \frac{K_I + K_P s + K_D s^2}{s(1+sT_F)},
\end{aligned} \tag{4}
$$

as shown in **Figure 2**, where parameters b and c are reference-weighting parameters for the proportional and derivative terms, respectively.

3. MOMI and DRMO tuning methods

The MOMI and DRMO methods, as mentioned earlier, are based on the Magnitude Optimum (MO) method, which goes back to Whitley in 1946 [10]. The MO method shapes the closed-loop amplitude frequency response equal to one in a wide frequency range [6, 7, 10, 12–14, 21]. Such a closed-loop frequency response is usually "mirrored" into a fast and stable closed-loop time response.

The calculation of controller parameters has been simplified when using the MO method by determining the process characteristic areas or moments, which can be measured directly from the time responses during the change of the process steady-state [12, 15, 21, 22]. The mentioned areas or moments (A_1 to A_k) can also be calculated from the process model:

$$
\begin{aligned}
A_0 &= K_{PR} \\
A_1 &= K_{PR}(a_1 - b_1 + T_{del}) \\
A_2 &= K_{PR}\left[b_2 - a_2 - b_1 T_{del} + \frac{T_{del}^2}{2!}\right] + A_1 a_1 \\
&\vdots \\
A_k &= K_{PR}\left[(-1)^{k+1}(a_k - b_k) + \sum_{i=1}^{k}(-1)^{k+i}\frac{T_{del}^i b_{k-i}}{i!}\right] + \sum_{i=1}^{k-1}(-1)^{k+i-1}A_i a_{k-i}
\end{aligned} \tag{5}
$$

The controller parameters, for a given filter time constant T_F, are then calculated as follows:

$$
\begin{bmatrix} K_I \\ K_P \\ K_D \end{bmatrix} = \begin{bmatrix} -A_1^* & A_0^* & 0 \\ -A_3^* & A_2^* & -A_1^* \\ -A_5^* & A_4^* & -A_3^* \end{bmatrix}^{-1} \begin{bmatrix} -0.5 \\ 0 \\ 0 \end{bmatrix}, \tag{6}
$$

where the modified areas A_0^* to A_5^* are:

$$
\begin{aligned}
A_0^* &= A_0 \\
A_1^* &= A_1 + A_0 T_F \\
A_2^* &= A_2 + A_1 T_F + A_0 T_F^2 \\
&\vdots
\end{aligned} \tag{7}
$$

The reference-weighting factors are $b = c = 1$. Note that the areas (moments) in expression (6) apply areas of the process including the controller filter G_F with time constant T_F (4):

$$G_F = \frac{1}{(1 + sT_F)} \tag{8}$$

by using expression (7) [9]. The aforementioned modification of the method, referred to as the MOMI method, allowed the controller parameters to be computed directly from the process time response [12, 21] or from the process transfer function.

Since the MOMI method aims at optimising the tracking performance, the disturbance rejection performance may be degraded for some types of processes.

To improve the disturbance-rejection performance, the optimisation criteria of the MOMI method were modified accordingly. The new method, referred to as the DRMO (Disturbance-Rejection-Magnitude-Optimum) method, achieved significantly improved disturbance rejection performance [9, 16, 17].

Similar to the MOMI method, the controller parameters in the DRMO method are also based on characteristic areas or moments. Therefore, the controller parameters can be calculated either from the process time-response or from the process transfer function.

The PID controller parameters are calculated according to the following expressions when using the DRMO method [9, 16, 17]:

$$K_P = \frac{\beta - \sqrt{\beta^2 - \alpha\gamma}}{\alpha}$$
$$K_I = \frac{\left(1 + K_P A_0^*\right)^2}{2\left(K_D A_0^{*2} + A_1^*\right)} \tag{9}$$

where

$$\alpha = A_1^{*3} + A_0^{*2} A_3^* - 2A_0^* A_1^* A_2^*$$
$$\beta = A_1^* A_2^* - A_0^* A_3^* + 2K_D\left(A_0^* A_1^{*2} - A_0^{*2} A_2^*\right)$$
$$\gamma = K_D^3 A_0^{*4} + 3K_D^2 A_0^{*2} A_1^* + K_D\left(2A_0^* A_2^* + A_1^{*2}\right) + A_3^* \tag{10}$$

and the derivative gain K_D is calculated directly from expression (6). The reference-weighting factors are $b = c = 0$.

The DRMO tuning method significantly improved the disturbance rejection performance, especially for the lower-order processes. However, the reference tracking becomes slower due to the reference-weighting factors $b = c = 0$ in the 2-DOF control structure (4). The problem can be circumvented by including a simple disturbance estimator in the control scheme. Such a solution is denoted as DE-MOMI method.

4. DE-MOMI tuning method

In order to improve the disturbance rejection response, while retaining the tracking response obtained by the MOMI method, a disturbance estimator has been added to the PID controller $G_C(s)$ (2), as depicted in **Figure 3**.

The disturbance estimator consists of the process model G_M, the inverse process model G_{MI} and the filter G_{FD}. In hypothetical case, when the process model is ideal representation of the bi-proper process without time-delay, and the filter $G_{FD} = 1$:

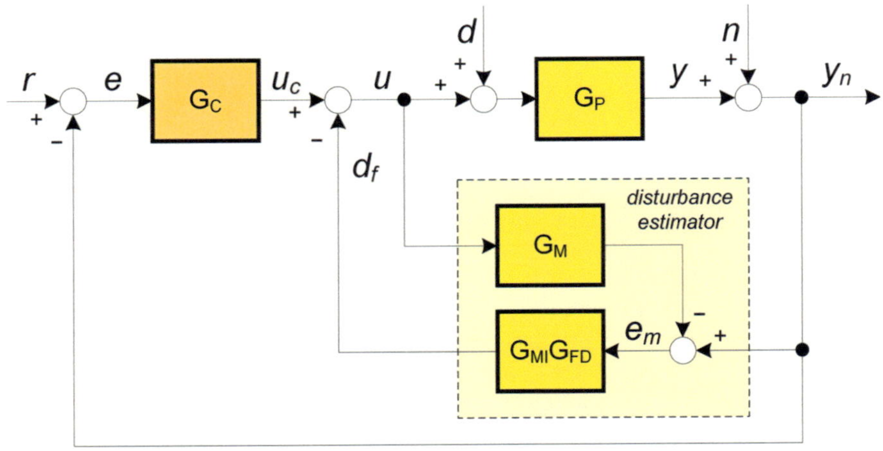

Figure 3.
The PID controller with disturbance estimator.

$$G_M = G_P, G_{MI} = G_M^{-1}, G_{FD} = 1, \quad (11)$$

the estimated disturbance d_f equals to the actual disturbance d:

$$d_f = d. \quad (12)$$

In this case the ideal disturbance compensation is achieved. However, in practice, model mismatch may occur (due to changing process characteristics in time or working point, lower-order process model or the process non-linearity), and the inverse of the process usually cannot be obtained, since majority of the actual processes are either strictly proper or they have time delays. Therefore, another strategy is required.

For practical applications, the solution has to be as simple as possible. In this manner we decided to use the following process model, the inverse process model and the disturbance estimator filter:

$$\begin{aligned} G_M &= \frac{K_{PRM}e^{-sT_{delm}}}{1 + a_{1m}s + a_{2m}s^2} \\ G_{MI} &= \frac{1 + a_{1m}s + a_{2m}s^2}{K_{PRM}} \\ G_{FD} &= \frac{K_{FD}}{(1 + sT_{FD})^3} \end{aligned} \quad (13)$$

where K_{PRM} and T_{delm} are the process model gain and time delay and a_{1m} and a_{2m} are the process model dynamic parameters. Parameters K_{FD} and T_{FD} are the disturbance filter gain and time constant, respectively.

The remaining question is how to obtain the process model if the actual process is of the higher order or if the actual process is not known (e.g. the areas (moments) were calculated directly from the process time-response)? Fortunately, the process model can be calculated directly from the obtained areas (5), as derived in [23]:

$$\frac{T_{delm}^3}{6} - \frac{A_1}{2A_0}T_{delm}^2 + \left(\frac{A_1^2}{A_0^2} - \frac{A_2}{A_0}\right)T_{delm} + \frac{2A_1A_2}{A_0^2} - \frac{A_3}{A_0} - \frac{A_1^3}{A_0^3} = 0$$

$$K_{PRM} = A_0$$

$$a_{1m} = \frac{A_1}{A_0} - T_{delm}$$

$$a_{2m} = -\frac{A_2}{A_0} + \frac{a_{1m}A_1}{A_0} + \frac{T_{delm}^2}{2} \tag{14}$$

The process model delay T_{delm} is calculated from the third-order equation in (14). The solution is the smallest real positive result [23].

Now, all the model parameters are known and the disturbance filter G_{FD} parameters should be derived. Before continuing the derivation we should be aware of the fact that $G_{MI}(s)$ is not proper, so it cannot be realised in practice without the accompanied filter $G_{FD}(s)$. Multiplication of both is strictly proper, so the entire block can be easily implemented inside the controller.

Derivation of disturbance filter parameters depends mainly on desired disturbance rejection performance. It is natural that the disturbance signal reconstruction (d_f) is faster if the filter time constant T_{FD} is smaller. In hypothetical case, if $G_M = G_P$, the reconstructed disturbance signal d_f becomes:

$$d_f = K_{FD}\frac{e^{-sT_{delm}}}{(1+sT_{FD})^3}d \tag{15}$$

With sufficiently small time constant ($T_{FD} \to 0$), where the disturbance filter gain K_{FD} = 1, and there is no process time delay, $df \approx d$. In this case the reconstructed disturbance signal d_f perfectly compensates the disturbance d. On the other hand, smaller disturbance filter time constant significantly increases the process output noise present in the measurements and forwards it to the controller output. Therefore, the T_{FD} should be selected according to the tolerated noise gain of the disturbance estimator, as will be discussed in detail in the next sub-chapter.

One remaining parameter of the disturbance filter G_{FD} (13) is the gain K_{FD}. One would, naturally, expect that the most optimal value should be K_{FD} = 1, since only in this case, after some time, d_f becomes the same to d (15). Therefore, the process input disturbance d is eliminated by the reconstructed disturbance d_f. However, as will be shown below, the optimal disturbance response is obtained with lower values of the gain K_{FD}. Namely, due to the disturbance compensator, the external process input signal d generates the delayed reconstructed disturbance signal d_f (15). Combined together, the actual process input u, due to disturbance d, is d-d_f. The step-like signal d, therefore, generates pulse-like actual process input disturbance signal d-d_f. Since the PID controller is present in the loop, and it contains the integrating term, the process output (y) deviation in one direction (e.g. above the reference) should be compensated by the process output deviation in the opposite direction (e.g. below the reference). Namely, when using K_{FD} = 1, the integral of the control error must be:

$$\int_{t=0}^{\infty} e(t)dt = 0. \tag{16}$$

It means that, by applying K_{FD} = 1, the additional process undershoot, after the initial process overshoot due to the disturbance d, is inevitable.

Figure 4 shows an example on delayed second-order process, when applying the step-wise external process input disturbance signal d, and when using K_{FD} = 1 (upper figure) and K_{FD} = 0.44 (lower figure). The process output undershoot in the upper figure is clearly seen. By appropriately reducing the filter gain to K_{FD} = 0.44, the disturbance rejection response is improved (lower figure).

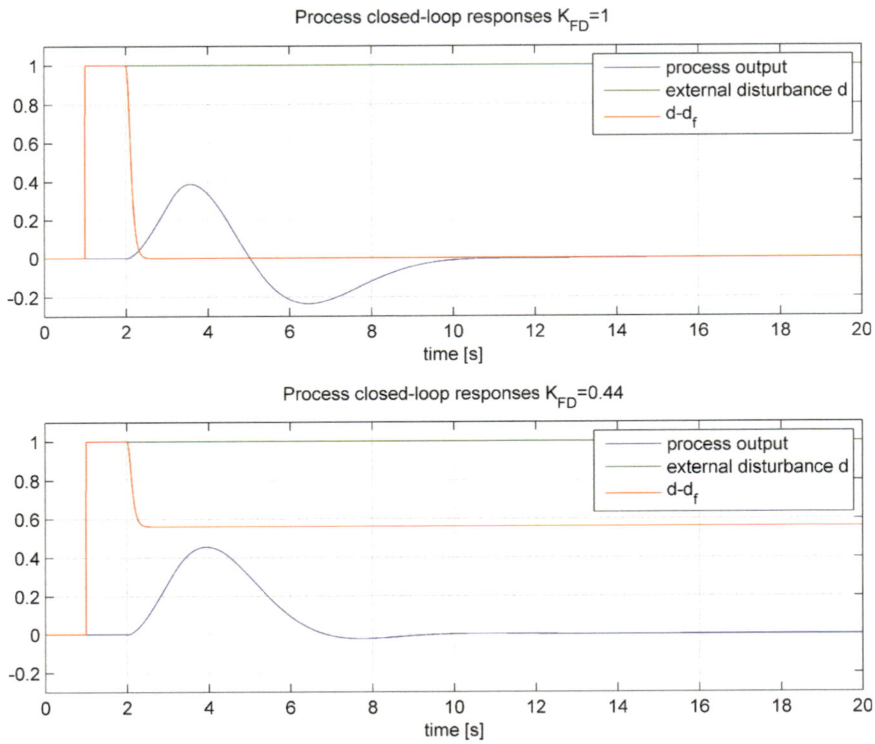

Figure 4.
The closed-loop signals when applying step-wise external process input disturbance signal with disturbance filter gains $K_{FD} = 1$ *(upper figure) and* $K_{FD} = 0.44$ *(lower figure).*

The remaining question is how to find the most appropriate filter gain K_{FD}. Certainly, the K_{FD} should be chosen so as that the disturbance rejection is optimised. Here we can use the same optimisation criteria as in the DRMO tuning method. Therefore, the transfer function $G_{CLD}(s)$ between the external disturbance (d) and the process output (y):

$$G_{CLD}(s) = \frac{Y(s)}{D(s)} = \frac{G_M\left(1 - G_{FD}e^{-sT_{delm}}\right)}{1 + G_M G_P} \tag{17}$$

should be optimised according to the modified MO criterion [9, 16]. Note that expression (17) holds when the process and the model transfer functions are equivalent. Since the disturbance filter time constant is defined, and all of the controller and the model parameters are calculated, the only optimisation parameter is the gain K_{FD}. By using similar derivation as in [9, 16], the optimal filter gain K_{FD} is calculated as:

$$K_{FD} = \frac{-b_0 + \sqrt{b_0^2 - 4a_0c_0}}{2a_0} \tag{18}$$

where

$$a_0 = -d_0 + 2f_0K_{IP} + K_{IP}^2\left(T_F^2 - 3T_{FD}^2\right)$$
$$b_0 = 2d_0 - 4f_0K_{IP} + K_{IP}^2\left(-2T_F^2 + 12T_{FD}^2 + 6T_{FD}T_{delm} + T_{delm}^2\right)$$
$$c_0 = -d_0 + 2f_0K_{IP} + K_{IP}^2T_F^2$$

$$K_{IP} = K_I K_{PRM}$$
$$K_{PP} = K_P K_{PRM}$$
$$K_{DP} = K_D K_{PRM}$$
$$d_0 = (1 + K_{PP})^2$$
$$f_0 = a_{1m} + K_{DP} + T_F + T_{delm} \qquad (19)$$

For the given controller filter (T_F) and the disturbance filter (T_{FD}) time constants (note that the calculation of both time constants, according to the desired level of noise, will be derived in the next sub-chapter), the calculation of remaining controller, model and disturbance filter parameters proceeds as given in **Figure 5**.

Illustrative example 1

To illustrate the proposed design of DE-MOMI method, according to control structure in **Figure 5**, let us calculate the controller, model and disturbance filter parameters for the following processes:

$$G_{P1} = \frac{e^{-0.5s}}{(1+s)^2}$$
$$G_{P2} = \frac{e^{-0.2s}}{(1+s)^3} \qquad (20)$$

The a-priori chosen filter time constants were:

$$T_F = T_{FD} = 0.1 \qquad (21)$$

The characteristic areas, calculated from (5) and (7), are given in **Table 1**.

Next, the PID controller parameters are calculated from (6) and from (9), since we are going to compare the proposed DE-MOMI method with MOMI and DRMO methods. The calculated controller parameters are given in **Table 2**.

The process models G_M and inverse process models G_{MI} are then calculated from (14), where G_{MI} is the inverse of G_M without time-delay:

1. Calculate the characteristic areas or moments from the process time-response [9,12] or from (5) if the process transfer function is known in advance.
2. Calculate modified areas according to the chosen T_F from (7).
3. Calculate PID controller parameters from (6).
4. Calculate process model parameters from (14).
5. Calculate the disturbance filter gain K_{FD} from (18).

Figure 5.
Calculation of the controller, model and filter parameters.

	A_0	A_1	A_2	A_3	A_4	A_5
Areas G_{P1}	1	2.50	4.13	5.77	7.42	9.07
Areas G_{P1} with controller filter	1	2.60	4.39	6.21	8.04	9.87
Areas G_{P2}	1	3.20	6.62	11.26	17.12	24.21
Areas G_{P2} with controller filter	1	3.30	6.95	11.96	18.32	26.04

Table 1.
The calculated areas for the processes (20) without and with the controller filter.

Controller parameters	K_P	K_I	K_D
MOMI controller for G_{P1}	1.81	0.89	0.93
DRMO controller for G_{P1}	2.25	1.49	0.93
MOMI controller for G_{P2}	1.61	0.64	1.08
DRMO controller for G_{P2}	1.93	0.98	1.08

Table 2.
The calculated controller parameters for the processes (20) for MOMI (6) and DRMO (9) method, taking into account the chosen controller filter T_F = 0.1.

$$G_{M1} = \frac{e^{-0.5s}}{1 + 2s + s^2}$$
$$G_{MI1} = 1 + 2s + s^2$$
$$G_{M2} = \frac{e^{-0.616s}}{1 + 2.58s + 1.84s^2} \tag{22}$$
$$G_{MI2} = 1 + 2.58s + 1.84s^2$$

Finally, the disturbance filter gain K_{FD}, when taking into account the chosen T_{FD} = 0.1, is then calculated from (18):

$$K_{FD1} = 0.57$$
$$K_{FD2} = 0.59 \tag{23}$$

Therefore, the complete inverse of the models with accompanying disturbance filters (see **Figure 3**) are the following:

$$G_{MI1}G_{FD1} = \frac{0.57(1 + 2s + s^2)}{(1 + 0.1s)^3}$$
$$G_{MI2}G_{FD2} = \frac{0.59(1 + 2.58s + 1.84s^2)}{(1 + 0.1s)^3} \tag{24}$$

The closed-loop responses, obtained with the calculated controller, model and filter parameters, for the MOMI, DRMO and the proposed DE-MOMI method, are given in **Figures 6** and 7. At t = 0 s, the reference value (r) was changed from 0 to 1 and at half of experiment time the process input disturbance (d) was changed from 0 to 1. It is obvious that the disturbance rejection performance of the DE-MOMI method is the best. Note that when applying the DE-MOMI method, due to the difference between the actual process and the process model in the second example (G_{P2}), the process input signal, during the reference change, is not smooth. This is expected, since the inverse process model with filter is amplifying the difference between the actual process and the process model. In this case, the response can be made smoother by increasing the disturbance filter time constant (T_{FD}). Note that a possible limitation of the control signal can also help to smooth out the oscillations after the reference step [24].

The disturbance rejection performance of the DE-MOMI method can be increased by decreasing the disturbance filter time constant T_{FD}. However, as already mentioned above, the process input signal can become oscillatory when the actual process and the process model differ. In this case, too small T_{FD} can even render the closed-loop system unstable. Besides that, the process noise (signal n in

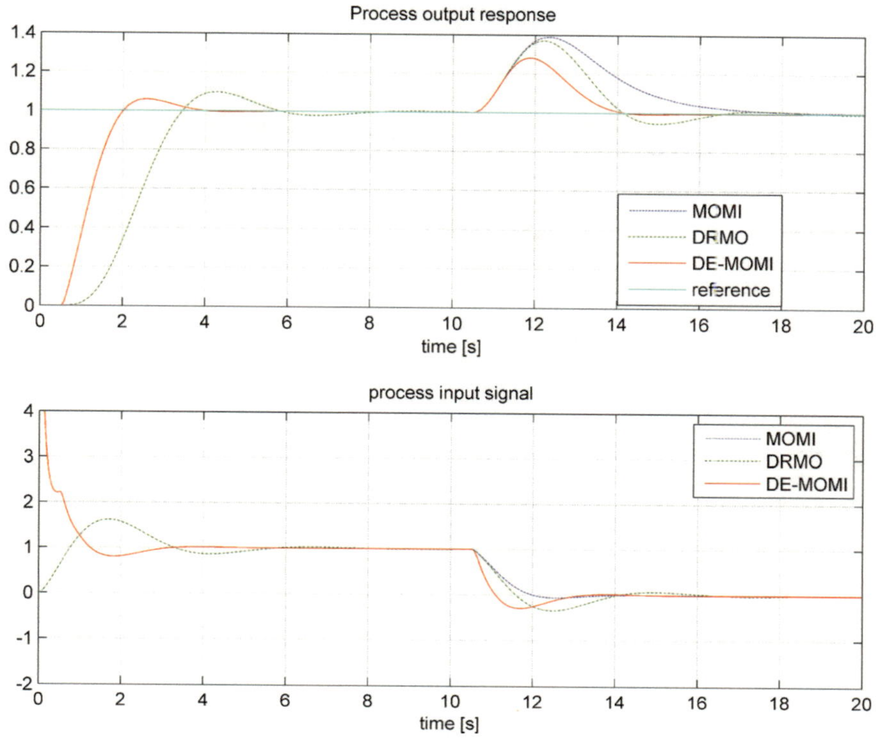

Figure 6.
The closed-loop responses on the process G_{P1}, *when using the MOMI, DRMO and DE-MOMI method.*

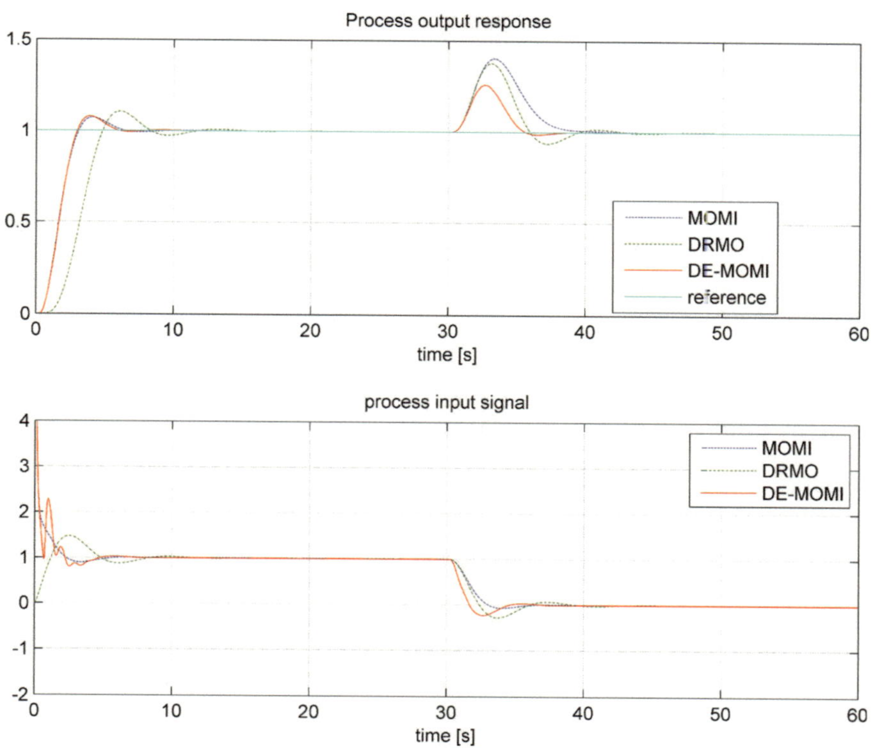

Figure 7.
The closed-loop responses on the process G_{P2}, *when using the MOMI, DRMO and DE-MOMI method.*

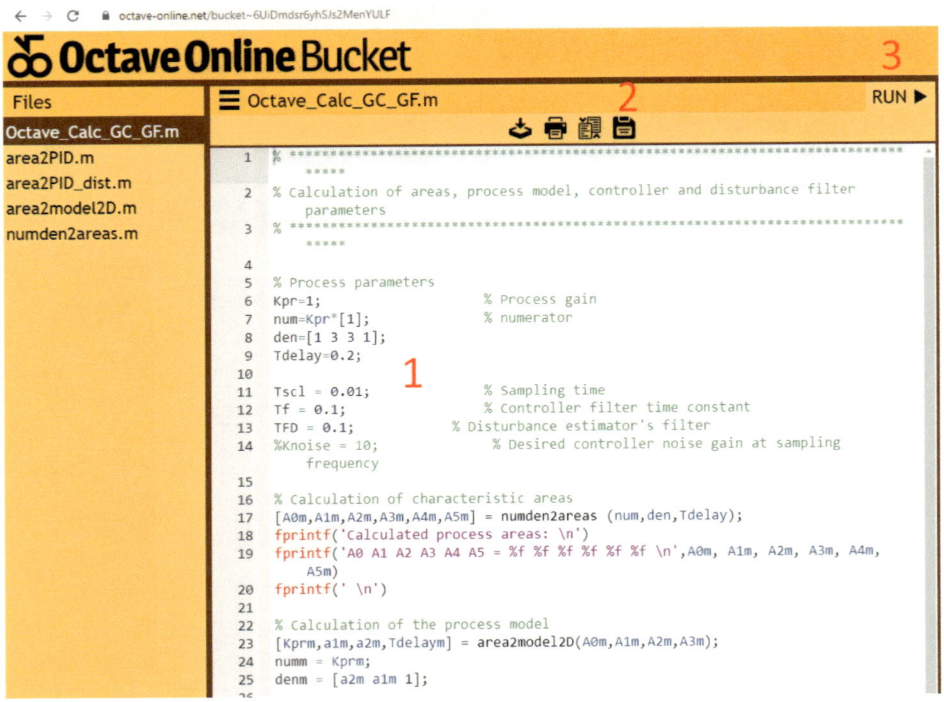

Figure 8.
The website layout for the calculation of the controller and the DE parameters.

Figure 3) is also amplified via block $G_{MI}G_{FD}$, so small T_{FD} can cause excessive noise of signal d_F. The selection of T_{FD} is, therefore, important in practical realisation of the DE-MOMI method.

Calculating the controller and DE parameters is a relatively simple process. However, to simplify it even further, all Matlab/Octave scripts are available on the OctaveOnline Bucket website [25]. The layout of the website is shown in **Figure 8**. To calculate the controller and DE parameters, the user must 1) change the process and filter parameters, 2) press the "Save" button, and 3) press the "Run" button. The script will be executed and on the right side of the web screen all calculated parameters will be displayed. Note that users can change the content of the script only temporarily.

5. Noise attenuation of DE-MOMI method

As already mentioned in the previous sub-chapter, the output noise of disturbance estimator (d_F) depends on the selection of disturbance filter T_{FD}. However, according to **Figure 3**, some noise is also present at the output of the PID controller block (signal u_C). In this sub-chapter we will give some guidelines regarding the noise attenuation in practical realisation of DE-MOMI controller.

In practice, it is important to keep the controller output noise within some limits. Namely, if the controller's and the estimator's filter time constants are too low, the DE-MOMI controller output noise can be so high that the controller would be useless in practice.

The controller noise is mainly caused by the process output noise n (see **Figure 3**). The noise power at the controller output (u) depends on the power of measurement noise n and the frequency properties of noise, PID controller and disturbance estimator. The relation between the filters (T_F and T_{FD}) time constants and the controller output noise is rather complex, but can be calculated according to Parseval's theorem if the measurement noise frequency characteristics are known. However, this relation is higher-order and non-linear. Therefore, the search for adequate filter time constants T_F and T_{FD} would require optimisation procedure, which would significantly complicate the otherwise simple method.

In practice, on the other side, it is enough to keep the noise sufficiently low at some sufficiently high frequency. The definition of "high frequency" is arguable. In discrete-realisation of the controller, the sampling frequency is

$$f_S = \frac{1}{T_S} \tag{25}$$

where T_S is the controller's sampling time. The highest signal, which may be sent to discrete function is, due to Shannon's theorem, $f_S/2$. Therefore, any frequency close to $f_S/2$ can be considered as high frequency. In this research we have arbitrarily decided that the "high frequency" f_{HF} is the quarter of controller's sampling frequency f_S:

$$\begin{aligned} f_{HF} &= 0.25 f_S \\ \omega_{HF} &= 2\pi f_{HF} = \frac{0.5\pi}{T_S} \end{aligned} \tag{26}$$

As already mentioned above, the source of controller noise is the process output noise n (**Figure 3**). In DE-MOMI controller, the overall high-frequency control noise consists of the PID controller (u_{PIDn}) and the disturbance estimator (u_{DEn}) high-frequency noise:

$$\begin{aligned} u_{PIDn}(\omega_{HF}) &= K_{PIDn} n(\omega_{HF}) \\ u_{DEn}(\omega_{HF}) &= K_{DEn} n(\omega_{HF}), \end{aligned} \tag{27}$$

where K_{PIDn} and K_{DEn} are the high-frequency gains (around frequency ω_{HF}) of the PID controller and the disturbance estimator, respectively.

In practical applications of the DE-MOMI method, the noise specifications (limitations) should be given in as simple form as possible for the user (operator). We decided that the actual parameters, given by the user should be the high-frequency gains of the controller (K_{PIDn}) and the disturbance estimator (K_{DEn}). Therefore, in practice, by selecting the mentioned two gains, the user would limit the amount of controller noise at high frequencies.

The actual gain of the PID controller around the chosen high frequency ω_{HF} can be calculated from the controller transfer function (2):

$$K_{PIDn} = \frac{\sqrt{\left(K_I - K_D\omega_{HF}^2\right)^2 + K_P^2\omega_{HF}^2}}{\omega_{HF}\sqrt{1 + T_F^2\omega_{HF}^2}} \tag{28}$$

The controller filter time constant can then be calculated as:

$$T_F = \frac{\sqrt{\omega_{HF}^4 K_D^2 + \omega_{HF}^2\left(K_P^2 - 2K_I K_D - K_{PIDn}^2\right) + K_I^2}}{\omega_{HF}^2 K_{PIDn}} \tag{29}$$

Since the PID controller parameters depend on the filter time constant T_F, the T_F should be calculated by an iterative procedure given in **Figure 9**.

The calculation of the disturbance filter high-frequency gain K_{DEn} is similar as for the PID controller:

$$K_{DEn} = \sqrt{\frac{K_{FD}^2\left(1+\omega_{HF}^2\left(a_{1m}^2-2a_{2m}\right)+\omega_{HF}^4 a_{2m}^2\right)}{K_{PRM}^2\left(1+\omega_{HF}^2 T_{FD}^2\right)^3}} \tag{30}$$

In a similar manner, the disturbance filter time constant can be derived as:

$$T_{FD} = \frac{1}{\omega_{HF}}\sqrt{\sqrt[3]{\frac{K_{FD}^2\left(1+\omega_{HF}^2\left(a_{1m}^2-2a_{2m}\right)+\omega_{HF}^4 a_{2m}^2\right)}{K_{PRM}^2 K_{DEn}^2}}-1} \tag{31}$$

Since the calculated filter gain K_{FD} depends on the filter time constant T_{FD} (see expression (18)), the calculation of expression (31) is iterative as well, as given in **Figure 10**.

Illustrative example 2

Let us illustrate the calculation procedure for the following processes:

$$G_{P3} = \frac{e^{-0.2s}}{(1+0.2s)(1+s)}$$

$$G_{P4} = \frac{e^{-s}}{(1+s)^4} \tag{32}$$

Note that other process models were chosen as in the previous case (20) in order to test different types of processes. The chosen high-frequency gains of the PID controller and the disturbance filter are K_{PIDn} = 4 and K_{DEn} = 4, respectively. The chosen closed-loop sampling time was T_S = 0.01 s. Therefore, the chosen high-frequency is:

$$\omega_{HF} = \frac{0.5\pi}{T_S} = 157.1\,s^{-1} \tag{33}$$

1. Set the controller filter time constant to T_F=0 or to some small value.
2. Calculate the PID controller parameters from (6).
3. Calculate T_F from (29), according to the desired high-frequency gain K_{PIDn}.
4. Repeat steps 2 and 3 few times (about 2-3 iterations are usually sufficient)

Figure 9.
Calculation of the filter and controller parameters according to the desired controller high-frequency gain K_{PIDn}.

1. Set the controller filter time constant to T_{FD}=0 or to some small value.
2. Calculate the K_{FD} gain from (18).
3. Calculate T_{FD} from (31), according to the desired high-frequency gain K_{DEn}.
4. Repeat steps 2 and 3 few times (about 2-3 iterations are usually sufficient)

Figure 10.
Calculation of the disturbance filter parameters according to the desired disturbance filter high-frequency gain K_{DEn}.

The initially chosen filter time constants were (the values are not critical):

$$T_F = T_{FD} = 0.1\,s \tag{34}$$

The characteristic areas are calculated from (5). For the given high-frequency gain K_{PIDn} = 4, the filter and controller parameters are calculated according to procedure given in **Figure 9**. The calculated filter time constants (after 2 iterations) were

$$\begin{aligned} T_{F3} &= 0.119\,s \\ T_{F4} &= 0.192\,s \end{aligned} \tag{35}$$

Note that indexes 3 and 4 in above filter time constants stand for the processes G_{P3} and G_{P4}, respectively.

The areas are given in **Table 3** and the controller parameters are given in **Table 4**.

The process models G_M and inversed process models G_{MI} are then calculated from (14):

$$\begin{aligned} G_{M3} &= \frac{e^{-0.2s}}{1+1.2s+0.2s^2} \\ G_{MI3} &= 1+1.2s+0.2s^2 \\ G_{M4} &= \frac{e^{-1.94s}}{1+3.06s+2.69s^2} \\ G_{MI4} &= 1+3.06s+2.69s^2 \end{aligned} \tag{36}$$

According to the chosen high-frequency gain K_{DEn} = 4, the T_{FD} and K_{FD} were calculated according to the procedure given in **Figure 10** (2 iterations were sufficient):

$$\begin{aligned} T_{FD3} &= 0.06 \\ T_{FD4} &= 0.116 \\ K_{FD3} &= 0.69 \\ K_{FD4} &= 0.36 \end{aligned} \tag{37}$$

Therefore, the complete inverse of the models with accompanying disturbance filters (see **Figure 3**) are the following:

$$G_{MI3}G_{FD3} = \frac{0.69(1+1.2s+0.2s^2)}{(1+0.06s)^3}$$

	A_0	A_1	A_2	A_3	A_4	A_5
Areas G_{P3}	1	1.40	1.50	1.52	1.53	1.53
Areas G_{P3} with controller filter	1	1.52	1.68	1.72	1.73	1.73
Areas G_{P4}	1	5.00	14.50	32.17	60.71	102.8
Areas G_{P4} with controller filter	1	5.19	15.50	35.14	67.45	115.8

Table 3.
The calculated areas for the processes (32) without and with the controller filter.

Controller parameters	K_P	K_I	K_D
MOMI controller for G_{P3}	2.35	1.88	0.48
DRMO controller for G_{P3}	2.91	3.83	0.48
MOMI controller for G_{P4}	0.84	0.26	0.77
DRMO controller for G_{P4}	0.94	0.32	0.77

Table 4.
The calculated controller parameters for the processes (20) for MOMI (6) and DRMO (9) method, taking into account the calculated controller filters.

$$G_{MI4}G_{FD4} = \frac{0.36(1 + 3.06s + 2.69s^2)}{(1 + 0.116s)^3} \tag{38}$$

The closed-loop responses for the MOMI, DRMO and the proposed DE-MOMI method, are given in **Figures 11** and **12**. Again, the disturbance rejection performance of the DE-MOMI method is the best (note that the unity-step process input disturbance signal was applied at the half of experiment time). The level of controller output (u) noise is close to the expected one taken into account that both, the PID controller (u_C) and the disturbance estimator output (d_F) noise should be 4-times higher than the measurement noise at high frequencies.

The disturbance rejection performance of the DE-MOMI method can be additionally improved by increasing the high-frequency gain K_{DEn}. However, increased gain is associated with higher controller output noise and decreased closed-loop stability if the actual process and the process model differ.

The computation of the controller and the DE parameters can be performed similarly as before on another OctaveOnline Bucket website [26]. The calculation of

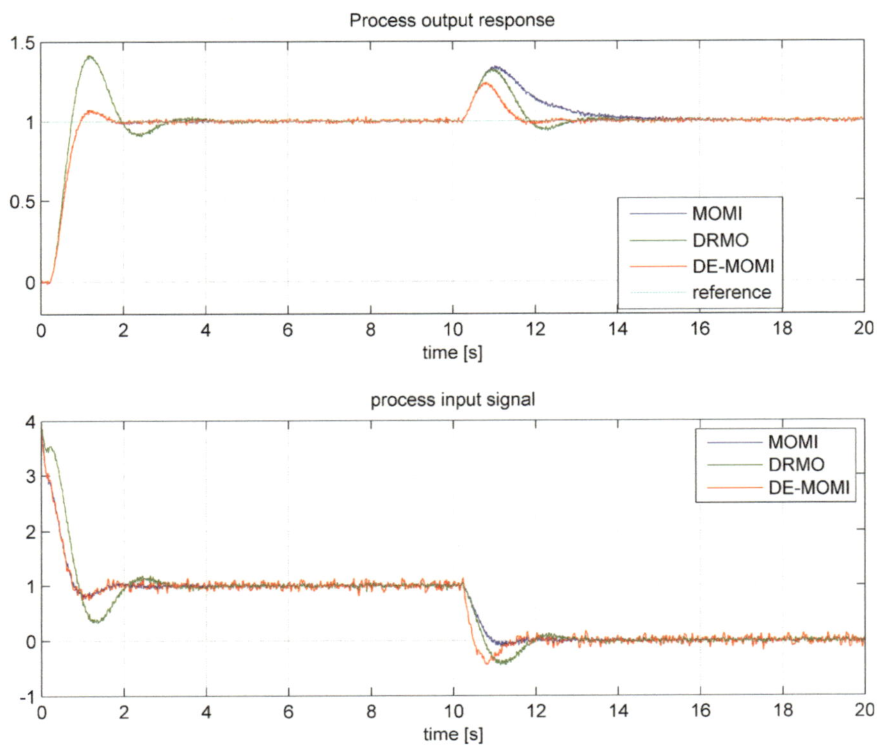

Figure 11.
The closed-loop responses on the process G_{P3}, when using the MOMI, DRMO and DE-MOMI method.

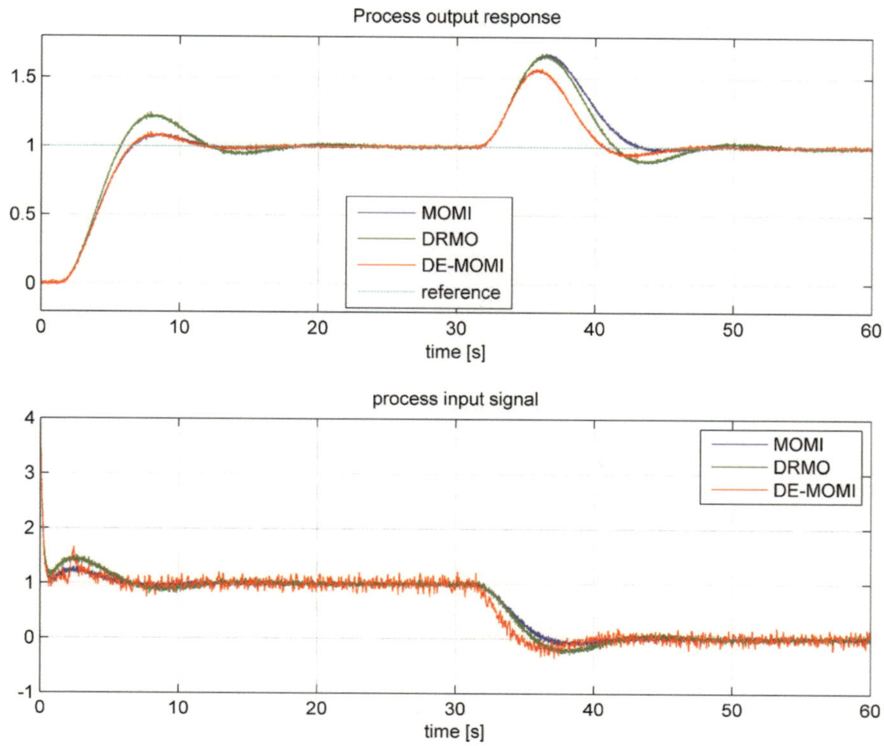

Figure 12.
The closed-loop responses on the process G_{P4}*, when using the MOMI, DRMO and DE-MOMI method.*

the parameters can be performed similarly as shown in **Figure 8**, with the difference that the name of the script is now Octave_Calc_GC_GF_Noise.m. To calculate the controller and DE parameters, the user must 1) change the process and noise gain parameters, 2) press the "Save" button, and then 3) press the "Run" button. The script will run and the right side of the web screen will display all the calculated parameters. Note that users can only temporarily change the contents of the script.

6. Comparison to some other methods

In this sub-chapter the proposed method will be compared to some other tuning methods based on non-parametric description of the process. Besides the already introduced MOMI and DRMO methods, the DE-MOMI method will be compared to Åström and Hägglund's tuning method [1] (denoted as "AH") and to ADRC method [27].

The AH method [1] is based on the calculation of the maximum sensitivity index M_S, which is the inverse of the smallest open-loop Nyquist curve distance to the critical point (−1,0i). The method was developed for values M_S = 1.4 and M_S = 2. In this comparison we will use M_S = 2, since it gives better disturbance-rejection performance. However, even though the process transfer function does not need to be derived, the method requires the identification of the process steady-state gain and the inflexion point along with maximum slope of the process output signal during the step-change of the process input signal. Note that those parameters usually require manual measurements and cannot be easily performed by using automatic calculation. The AH method is using the PID controller structure with adjustable reference-weighting factor b, and by fixing factor c = 0 (**Figure 2**).

The ADRC method [27–31] is based on a simple controller with three gains associated with extended state-observer (ESO), as shown in **Figure 13**.

The method does not require the process transfer function. However, few user-defined parameters, like the observer speed, the desired settling time and the main controller gain K_C, should be defined by the user before calculating the rest of ADRC parameters. As shown in **Figure 13**, the ADRC method is using control structure which consists of an extended state observer (ESO) with three gains (β_1, β_2 and β_3) and three controller gains (K_C, K_P and K_D) [27].

Since ADRC method depends on three user-defined parameters, which, in great extent, determine the closed-loop performance, we were limited to the set of processes tested in [27]. Someone would argue that, by limiting our choice to the mentioned processes, we are favouring the ADRC method. However, it should be noted that in [27], the ADRC method was tested on 8 different processes, so the choice of processes was actually not significantly limited. In this regard, the following two processes have been selected:

$$G_{P5} = \frac{1}{(1+s)(1+0.2s)(1+0.04s)(1+0.008s)}$$

$$G_{P6} = \frac{e^{-5s}}{(1+s)^3} \tag{39}$$

The PID controller parameters for the MOMI, DRMO, DE-MOMI and AH methods are given in **Tables 5** and **6**. The ADRC controller parameters are given in **Table 7**. The chosen high frequency gains for the PID controller and disturbance estimator are $K_{PIDn} = K_{DEn} = 20$ for G_{P5} and $K_{PIDn} = K_{DEn} = 4$ for G_{P6}. The higher gains were chosen for G_{P5}, since the closed-loop tracking and control performance was substantially improved when using higher gains. Increasing the gains for G_{P6} above 4 did not significantly improve the performance.

The sampling time for G_{P5} is chosen as $T_S = 0.001$ s and for G_{P6} as $T_S = 0.01$ s.

The closed-loop process responses are given in **Figures 14** and **15**. In both experiments the unity-step process input disturbance signal was applied at the half of experiment time.

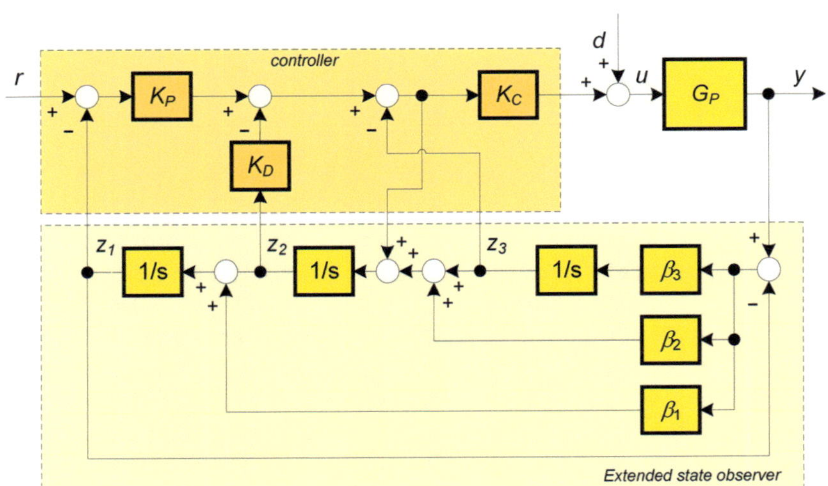

Figure 13.
The ADRC control structure with the controller gains (up) and the extended state observer (down).

Process	Tuning method	K_P	K_I	K_D	T_F	b	c
G_{P5}	MOMI	6.45	5.35	1.108	0.055	1	1
	DRMO	9.69	23.71	1.108	0.055	0	0
	DE-MOMI	6.45	5.35	1.108	0.055	1	1
	AH	21.35	53.05	2.22	0.055	0.24	0
G_{P6}	MOMI	0.53	0.126	0.66	0.165	1	1
	DRMO	0.57	0.140	0.66	0.165	0	0
	DE-MOMI	0.53	0.126	0.66	0.165	1	1
	AH	0.52	0.136	0.52	0.165	0.36	0

Table 5.
The calculated controller parameters for the processes (39) for MOMI, DRMO, DE-MOMI and AH method.

Process	K_{PRM}	a_{1m}	a_{2m}	T_{delm}	T_{FD}	K_{FD}
G_{P5}	1	1.205	0.205	0.043	0.018	0.909
G_{P6}	1	2.58	1.84	5.42	0.077	0.159

Table 6.
The calculated disturbance estimator's parameters for the processes (39) for DE-MOMI method.

Process	K_C	K_P	K_D	β_1	β_2	β_3
G_{P5}	1/5	100	20	120	4800	19200
G_{P6}	1/3	0.16	0.8	4.8	7.63	30.72

Table 7.
The calculated ADRC controller parameters for the processes (39).

It can be seen that the proposed DE-MOMI method, when compared to some other methods, gives quite good responses. The AH method for process G_{P5} gives somehow oscillatory response. For the same process, the ADRC method gives slightly oscillatory response during the reference change (see the process input signal). While DE_MOMI and MOMI methods clearly give the best tracking responses on process G_{P6}, all of the methods have similar disturbance-rejection performance. Only slightly oscillatory response can be observed for ADRC method.

For more objective comparison between the methods, the integral of absolute error (IAE) measure is used. The IAE value has been measured on tracking response (unity step-change of the reference r) and on disturbance rejection response (unity step-change of the process input disturbance d). The results are given in **Table 8**. It can be seen that the best values (marked with greyed colour) were obtained with DE-MOMI method.

The DE-MOMI method, therefore, compares favourably with few other methods, based on the non-parametric description of the process.

The process closed-loop responses for all the process models tested in this chapter (G_{P1} to G_{P6}) revealed that the proposed method can significantly improve the disturbance-rejection performance of the lower-order processes with smaller delays, while the improvement of the higher-order processes and/or processes with higher delays is not so significant. Therefore, the application of the method for lower-order processes with smaller delays might be beneficial in practice.

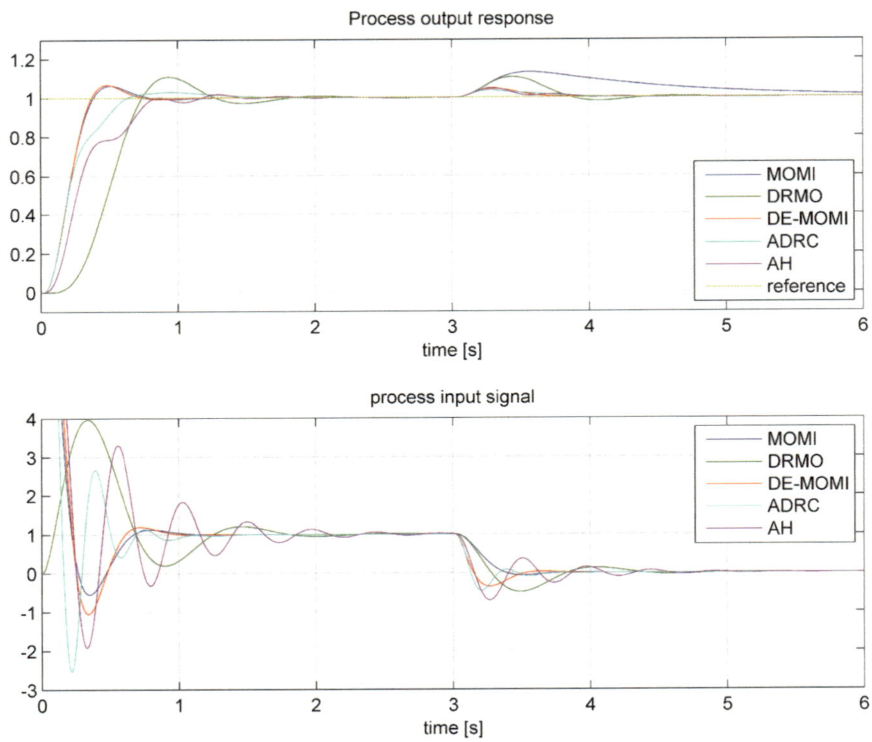

Figure 14.
The closed-loop responses on the process G_{P5}*, when using the MOMI, DRMO and DE-MOMI method.*

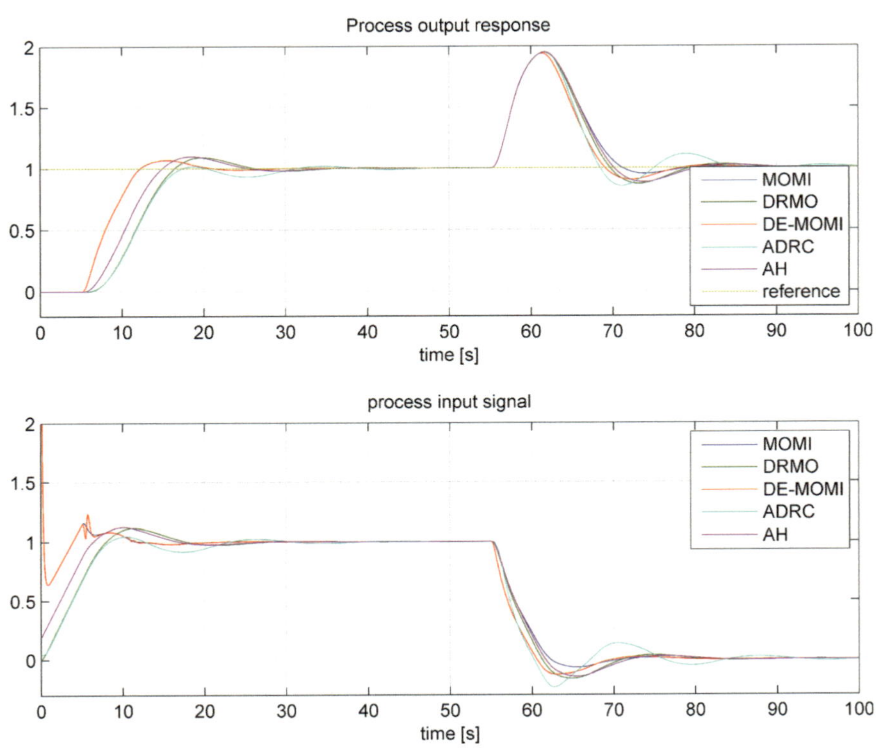

Figure 15.
The closed-loop responses on the process G_{P6}*, when using the MOMI, DRMO and DE-MOMI method.*

Process	experiment	DE-MOMI	MOMI	DRMO	AH	ADRC
G_{P5}	tracking	0.216	0.217	0.526	0.336	0.256
	DR	0.017	0.186	0.055	0.020	0.019
G_{P6}	tracking	8.66	8.66	12.34	11.06	12.32
	DR	7.80	8.42	8.79	8.89	8.83

Table 8.
The calculated IAE values for tracking and disturbance rejection (DR) responses for the processes (39).

7. Conclusions

In the chapter, it was shown that the disturbance rejection performance of the PID controller can be improved by adding a simple disturbance estimator (DE). The disturbance estimator consists of the process model and the inverse process model with DE filter. The advantage of the proposed approach is that the DE parameters can also be obtained directly from the nonparametric process data (time response of the process) without prior process identification. The same is true for the PID controller parameters, which are obtained using the MOMI tuning method. Of course, all PID and DE parameters can also be calculated from the process transfer function if it is known.

The proposed solution, called DE-MOMI method, has been tested on several different process models. It was shown that the control performance of the DE-MOMI method was significantly improved compared to similar MOMI and DRMO methods, especially for lower order processes with smaller time delays. In contrast, the improvements were noticeable but not as significant for higher order processes or processes with larger time delays. The additional advantage of the proposed method was that the tracking performance remained similar to that of the MOMI method.

The controller noise was controlled by the high frequency noise factors KPIDn and KDEn. The advantage of using these factors is that they can be easily understood and defined by the user.

The DE-MOMI method was also compared with some other non-parametric disturbance-rejection methods including the ADRC method. The results showed that the DE-MOMI method has either comparable or better control and tracking performance than the other tested methods. Nevertheless, it should be mentioned that the ADRC method uses a somewhat simpler control structure.

Future research activities could therefore focus on combining the advantages of the DE-MOMI and ADRC methods.

Acknowledgements

The authors gratefully acknowledge the contribution of the Ministry of Higher Education, Science and Technology of the Republic of Slovenia, Grant No. P2-0001 as well as the support by the grants APVV SK-IL-RD-18-0008 Platoon Modelling and Control for mixed autonomous and conventional vehicles: a laboratory experimental analysis and VEGA 1/0745/19 Control and modelling of mechatronic systems in emobility.

Author details

Damir Vrančić[1]* and Mikuláš Huba[2]

1 Jožef Stefan Institute, Department of Systems and Control, Ljubljana, Slovenia

2 Slovak University of Technology in Bratislava, Faculty of Electrical Engineering and Information Technology, Bratislava, Slovakia

*Address all correspondence to: damir.vrancic@ijs.si

References

[1] Åström, K.J. and Hägglund, T. (1995). PID controllers: Theory, Design, and Tuning, 2nd Edition. Instrument Society of America.

[2] Åström, K.J., Panagopoulos, H. and Hägglund, T. (1998). Design of PI Controllers based on Non-Convex Optimization. Automatica, 34 (5), pp. 585–601.

[3] Garcia, C. E. and Morari, M. (1982). Internal model control. A unifying review and some new results. Ind. Eng. Chem. Process Des. Dev. Vol. 21, pp. 308–323.

[4] O'Dwyer A. (2009). Handbook of PI and PID controller tuning rules; 3rd Edition. Imperial College Press.

[5] Freire H., Moura Oliveira P., Solteiro E. P. and Bessa M., (2014), Many-Objective PSO PID Controller Tuning, Controlo 2014, Porto, Portugal pp. 183-192.

[6] Gorez, R. (1997). A survey of PID auto-tuning methods. Journal A. Vol. 38, No. 1, pp. 3–10.

[7] Kessler, C. (1955). Über die Vorausberechnung optimal abgestimmter Regelkreise Teil III. Die optimale Einstellung des Reglers nach dem Betragsoptimum. Regelungstechnik, Jahrg. 3, pp. 40–49.

[8] Taguchi, H. and Araki, M. (2000). Two-degree-of-freedom PID controllers – their functions and optimal tuning. Proceedings of the IFAC Workshop on Digital Control (PID'00), Terrassa, 2000. pp. 95–100.

[9] Vrančić, D. (2011) Magnitude optimum techniques for PID controllers. In: PANDA, Rames C. (editor). Introduction to PID controllers: theory, tuning and application to frontiers areas. Rijeka: InTech, cop., pp. 75–102.

[10] Whiteley, A. L. (1946). Theory of servo systems, with particular reference to stabilization. The Journal of IEE, Part II, 93(34), pp. 353–372.

[11] Vrančić D., Strmčnik S., and Hanus R. (2000). Magnitude optimum tuning using non-parametric data in the frequency domain. PID'00 : preprints : IFAC workshop on digital control: past, present and future of PID control, Terrassa, Spain, April 5-7, 2000, pp. 438–443.

[12] Vrančić, D., Strmčnik, S. and Juričić, Đ. (2001). A magnitude optimum multiple integration method for filtered PID controller. Automatica. Vol. 37, pp. 1473–1479.

[13] Ba Hli, F. (1954). A General Method for Time Domain Network Synthesis. IRE Transactions – Circuit Theory, 1 (3), pp. 21–28.

[14] Hanus, R. (1975). Determination of controllers parameters in the frequency domain. Journal A, XVI (3).

[15] Strejc, V. (1960). Auswertung der dynamischen Eigenschaften von Regelstrecken bei gemessenen Ein- und Ausgangssignalen allgemeiner Art. Z. Messen, Steuern, Regeln, 3(1), pp. 7–10

[16] Vrančić D. Strmčnik S., Kocijan J. and Moura Oliveira P. B. de. (2010). Improving disturbance rejection of PID controllers by means of the magnitude optimum method. ISA transactions, ISSN 0019–0578, vol. 49, no. 1, pp. 47–56.

[17] Vrančić D., Kocijan J. and Strmčnik S. (2004). Simplified disturbance rejection tuning method for PID controllers. ASCC The 5th Asian Control Conference, July 20–23, Melbourne, Australia. Conference proceedings.

[18] Vrančić D., Oliveira P.M., Cvejn J. (2017) The Model-Based Disturbance Rejection with MOMI Tuning Method for PID Controllers. In: Garrido P., Soares F., Moreira A. (eds) CONTROLO 2016. Lecture Notes in Electrical Engineering, Vol 402. Springer, Cham. https://doi.org/10.1007/978-3-319-43671-5_8.

[19] Vrančić D., Moura Oliveira P. and Huba M. (2018) "Optimizing Disturbance Rejection by Using Model-Based Compensator with User-Defined High-Frequency Gains," 13th APCA International Conference on Automatic Control and Soft Computing (CONTROLO), Ponta Delgada, 2018, pp. 330–335, doi: 10.1109/CONTROLO.2018.8514546.

[20] Moura Oliveira P., Vrančić D. and Freire H., (2014), Dual Mode Feedforward-Feedback Control System, Controlo 2014, Porto, Portugal pp. 241-50

[21] Preuss, H. P. (1991). Model-free PID-controller design by means of the method of gain optimum (in German). Automatisierungstechnik, Vol. 39, pp. 15–22.

[22] Rake, H. (1987). Identification: Transient- and frequency-response methods. In M. G. Singh (Ed.), Systems & control encyclopedia; Theory, technology, applications. Oxford: Pergamon Press.

[23] Vrečko D., Vrančić D., Juričić Đ. and Strmčnik S. (2001). A new modified Smith predictor: the concept, design and tuning. ISA transactions, ISSN 0019–0578, vol. 40, pp. 111–121.

[24] Huba, M. (2006). Constrained pole assignment control. Current Trends in Nonlinear Systems and Control, L. Menini, L. Zaccarian, Ch. T. Abdallah, Edts., Boston: Birkhäuser, pp. 163–183.

[25] Vrančić, D. (2020). Matlab/Octave function Octave_Calc_GC_GF.m for the calculation of controller and GE parameters based on a given filter time constants. Web page: https://octave-online.net/bucket~6UiDmdsr6yhSJs2MenYULF.

[26] Vrančić, D. (2020). Matlab/Octave function Octave_Calc_GC_GF_Noise.m for the calculation of controller and GE parameters based on a given noise gains. Web page: https://octave-online.net/bucket~Hofor19mDWWkAzZrhgeG7q.

[27] Chen X., Li D., Gao Z. and Wang C. (2011). Tuning Method for Second-order Active Disturbance Rejection Control. Proceedings of the 30th Chinese Control Conference, July 22–24, 2011, Yantai, China, Conference proceedings.

[28] Gao Z., Huang Y., and Han J. (2001). An Alternative Paradigm For Control System Design. Proceedings of the 40th IEEE Conference on Decision and Control, Orlando, USA, December 4-7, 2001, Vol. 5, doi: 10.1109/CDC.2001.980926, pp. 4578–4585.

[29] Han J. (2009). From PID to Active Disturbance Rejection Control. IEEE transactions on industrial electronics, ISSN 0278–0046, vol. 56, no. 3, pp. 900–906.

[30] Huba, M., Moura Oliveira P., Vrančić D., and Bistak P. (2019). ADRC as an exercise for modeling and control design in the state-space. 6th International Conference on Control, Decision and Information Technologies, CoDIt 2019, Paris, France April 23–26, 2019, pp. 464–469. doi: 10.1109/CoDIT.2019.8820543.

[31] Wang W. (2012). The New Design Strategy on PID Controllers. In: VAGIA, M. (editor). PID Controller Design Approaches, ISBN 978–953–51-6186-8, pp. 229–252.

Chapter 4

Adjustment of the PID Gains Vector Due to Parametric Variations in the Plant Model in Terms of Internal Product

José Pinheiro de Moura and João Viana da Fonseca Neto

Abstract

The tuning of the gains of a controller with proportional-integral-derivative (PID) actions has been prevalent in the industry. The adjustment of these gains in PID controllers is often determined by classical methods, such as Ziegler-Nichols, and trial and error. However, these methods fail to deliver satisfactory performance and often do not meet specific project demands because of the inherent complexity of industrial processes, such as plant parameter variations. To solve the tuning problem in highly complex industrial processes, a controller adjustment method based on the internal product of PID terms is proposed, and a propagation matrix (PM) is generated by the numerator coefficients of the plant transfer function (TF). In the proposed method, each term of the PID controller is influenced by each of the numerator and the denominator coefficients. Mathematical models of practical plants, such as unloading and resumption of bulk solids by car dumpers and bucket wheel resumption, were employed to evaluate the proposed method. The obtained results demonstrated an assertive improvement in the adjustment gains from PID actions, thereby validating it as a promising alternative to conventional methods.

Keywords: parameter variations, industrial processes, internal product, propagation matrix, PID actions

1. Introduction

The tuning of the parameters of PID controllers is challenging and requires expertise to achieve superior performance [1]. PID controllers are extensively used in the industries. However, the controllers are often implemented without a derivative action because of the highly sensitive tuning of parameters, which affects the efficiency of the controller [2]. This study presents a methodology for tuning three terms of the PID controller simultaneously to ensure overall efficiency of the controller [3].

The advantage of the PID controller tuning methodology, which is based on the internal product of the PID terms that generates the propagation matrix (PM), is that a vector of the specified parameters of a characteristic polynomial can be projected, and an error vector is obtained on comparison with the parameters of the characteristic plant polynomial [4, 5]. The method minimizes the error from the specified parameters, thereby facilitating the design of a high-performance PID controller. The method also enables the allocation of the poles by direct replacement using specified parameters, thereby ensuring the desired operating point of the control system.

This chapter presents a formulation proposal to resolve the PID controller tuning problem. The proposal is based on the dot product of the gain vector parameters of the controller and the rows of the propagation matrix. The dot product represents the changes in the behavior of the plant that are determined by the parametric variations in the coefficients of the TF polynomial characteristic.

The following topics and proposal development are presented in the remainder of this chapter. In Section 2, a preliminary on the transfer functions of the plant and PID controller are presented. In terms of internal product, the main properties of PID controllers and the development of proposed method are presented in Section 3. Taking into account three industrial plants of the mining sector, computational evaluation experiments of the PID tuning proposal are presented in Section 4. Finally, the conclusion of the work is presented in Section 5.

2. Preliminaries

Adjusting the PID controller gain parameters is not a trivial task and requires in-depth knowledge from the experts. In this work, the problem of tuning PID controllers is based on original studies regarding polynomial compensators and problems strongly related to the specification of parameters to meet operational constraints in plant dynamics, presented as a particular form of compensators in the s domain [6, 7].

2.1 Mathematical model of the plant in terms of transfer function

The plant's dynamic system is represented by ordinary differential equations (ODE), described by TFs in the s domain (Laplace transform). The ODE concept in terms of TFs established in this chapter is in accordance with the block diagram shown in **Figure 1**, where the closed loop system relates the input and output signals: $R(s)$ is reference input, $W(s)$ is disturbance signal and $V(s)$ is noise signal, $Y(s)$ is plant output and $Y_m(s)$ is plant output measured by the sensor, U(s) and E(s) are the control effort and closed loop error, respectively, that are internal control system performance variables.

Applying the Laplace transforms to the control elements of **Figure 1**, the generalized TF with a polynomial structure in the s domain is obtained that is given by

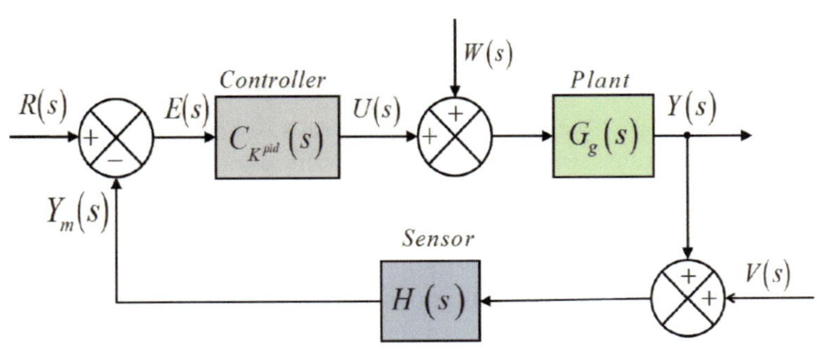

Figure 1.
Canonical block diagram of the closed-loop control system.

$$G_g(s) = \frac{b_o s^m + b_1 s^{m-1} + \ldots + b_{m-1} s + b_m}{a_o s^n + a_1 s^{n-1} + \ldots + a_{n-1} + a_n}, \tag{1}$$

where $G_g(s)$ is the general TF of the control system blocks diagram, related to **Figure 1**, n is the order of the plant model and the number of poles that are entered into the system and m is the number of zeros, which is associated with the PID actions of the controller. As an imposition of the controller gains values, the coefficients a_i and b_k are the adjustable parameters to compensate for the parametric variations of the plant. In the proposed formulation, the b_k coefficients are kept constant and adjustment are made only to the a_i parameters.

2.2 PID controller model

The controller model associated with the TF given in Eq. (1) is customized to perform the actions of the controller's PID terms, where $n = 1$, $m = 2$, $a_0 = 1$, the PID actions are represented by the transfer function that is given by

$$C_{K^{pid}}(s) = \frac{K_D s^2 + K_P s + K_I}{s}, \tag{2}$$

where $C_{K^{pid}}(s)$ is the controller model associated with the TF given in Eq. (1).

Adjustments of parameters that meet the project specifications, can be found in a large number of scientific and technical publications in controle specialized books, conferences and high quality journals [8]. The importance of developing methods for adjusting parameters of PID controllers and systematizing applications in industrial processes of real-world plants, has the objective of meeting the project specifications contained in technological advances, in order to guarantee the optimal adjustment of the parameters of the PID term of the controller [9, 10]. The challenge of tuning with optimal performance of the parameters of a PID controller, started around 1920 and continues to the present days [11–13].

The parameters of the PID controllers are adjusted to adapt to the tuning needs in a combination of proportionality associated with the proportional action, lead associated with the derivative action, and delay associated with the integral action of the error signal. However, there are still many problems that can be solved with computational intelligence-based algorithms. The purpose of this work is to contribute with a method of tuning PID controllers, which can support the development of electronic devices that contribute to technological advancement and the evolution of industry 4.0 with logical planning units, for optimal, robust decision-making and adaptability [14, 15]. Such units must be based on digital control technologies and embedded systems [16] in real time [17], to be reliably deployed in real-world systems [18].

To meet the demands of design specitifications, the proposed solution contributes to the evolution in approaches of optimal and adaptive control, providing the optimization of the figures of merit [19], ensuring a solution with satisfactory performance, meeting the requirements specified in projects, in a way that minimizes efforts of computational cost and control.

TF is specified in the factored form, that is, by the roots of the numerator and denominator polynomials associated with Eq. (1). TF in the factored form is represented in terms of product, where the designer inserts the specified or desired parameters. TF in the form of a product is given by

$$G_g(s) = K \frac{\prod_{k=1}^{m} (s - s_{zk})}{\prod_{i=1}^{n} (s - s_{pi})}, \tag{3}$$

where $s_{pi} = \omega_{di}$ are the poles and $s_{zk} = \sigma_k \omega_{dk}$ are the zeros of the dynamic system. The poles and zeros of the system are represented by the pair (ζ, ω_n), the first component is the damping factor and the second is the undamped natural frequency, $\sigma_i = \zeta_i \omega_{ni}$, and $\omega_{di} = \omega_{ni}\sqrt{1-\zeta_i^2}$.

3. PID adjustment problem

The PID controller adjustment problem is formulated based on the parametric difference between the specified coefficients and the original coefficients of the TF denominator polynomial. This formulation is based on the references [20, 21]. Where the authors present the development of models for optimized online optimization that is based on computational intelligence approaches.

The formulation of the proposed PID controller adjustment problem is presented in this section and is illustrated by the block diagram of **Figure 1**. Thus, in the context of the proposal, the performance matrix of the control system provides the means to determine the values of the parameters of the gain vector of the PID controller K^{pid}.

The models are represented in the form of internal produc <,>, which is a notation widely used in this text as the product of two vectors (internal product) and the models are called internal product models of the plant. The internal product is the appropriate form for analysis, allowing the designer to observe the impact of the earnings vector parameters K^{pid} (K_D, K_P and K_I), in the output of the dynamic system associated with the polynomial coefficients of the TF denominator.

The PID controller model, in terms of the ODE equations and the Laplace transform, obtains the TF of the PID controller in terms of the dot product, which is given by

$$C_{K^{pid}}(s) = \frac{\langle K^{pid}, s^{pid} \rangle}{s}, \tag{4}$$

$$K^{pid} = [K_D \; K_P \; K_I]^T, \tag{5}$$

and s^{pid} is vector-powered in s of the PID gains associated with transfer function numerator that is given by

$$s^{pid} = \left[s^2 \; s^1 \; s^0\right]^T. \tag{6}$$

3.1 PID model in the form of internal product

Inserting the characteristics of the plant, through the values of the coefficients of the polynomials of the numerator and the denominator (poles and zeros), associated with the mathematical models in terms of TFs given in Eq. (1) in internal product form $\langle \cdot \rangle$ and in Eq. (3) in generic form, which is given by

$$G_p(s) = K \frac{\langle b_k, s^m \rangle}{s^n + \langle a_i, s^{n-1} \rangle}, \tag{7}$$

where $G_p(s)$ is the TF in the form of internal product and the coefficients a_i with $i = 1, \ldots, n$ and b_k with $k = 0, \ldots, m$ are a combination of the s_{pi} poles and s_{zk} zeros.

3.2 Open-loop transfer function

The open-loop FT or direct branch of the control system is given by

$$G_p^{OL}(s) = K\frac{\langle K^{pid}, s^{pid}\rangle\langle b_k, s^m\rangle}{s^{n+1} + s\langle a_i, s^{n-1}\rangle}, \tag{8}$$

where $G_p^{OL}(s)$ is the TF of the plant in the open loop and K the gain of the plant and $n > m$.

The structure of TF is determined by the relationship $n+1 \geq 2+m$. For $n+1 = 2+m$, the system is proper and for $n+1 < 2+m$ the system is strictly proper, thereby establishing a general relationship between the order of the PID controller and the order of plant dynamics. This relationship ensures that the system structure is adequate, not allowing the system to present a nonpractical structure. In this way, it establishes that the relationship of the closed loop system is given by

$$n_{cl} = m^{PID} + n, \tag{9}$$

where m^{PID} can only assume 0 (zero) or 1 (one) values. The PID is observed to impose a proper TF, if the closed-loop system is of order $n+1$ and the numerator is of order $(m^{PID} + m)$.

According to the block diagram of **Figure 1**, the TFs $Y(s)/R(s)$, $W(s) = 0$, and $V(s) = 0$ are given by

$$G_p^{CL}(s) = \frac{C_{K^{pid}}(s)G(s)}{1 + C_{K^{pid}}(s)G(s)H(s)}. \tag{10}$$

3.3 Propagation of PID terms x b_k coefficients

The development of the polynomials of the numerator (zeros) and the denominator (poles) consists of the propagation of the gain vector K^{pid} of the controller by the numerator coefficients (b_k) associated with the coefficients of the denominated (a_i) TF of the plant. The equationing of the problem is given in the form of an internal product that weights the coefficients of the polynomial of zeros in the closed-loop and additive to the dynamics of the closed-loop transfer function.

3.3.1 Polynomial of zeros

When replacing Eqs. (1) and (2) in Eq. (10), the numerator polynomial of the closed-loop TF is obtained, which is given by

$$N^{CL}(s) = C_{K^{pid}}(s)G(s). \tag{11}$$

Expanding and ordering Eq. (11), one obtains

$$\begin{aligned} N^{CL}(s) = \ & (b_0K_D + b_{-1}K_P + b_{-2}K_I)s^{m+m^{pid}} \\ & + (b_1K_D + b_0K_P + b_{-1}K_I)s^{m+m^{pid}-1} \\ & + (b_2K_D + b_1K_P + b_0K_I)s^m \\ & + (b_3K_D + b_2K_P + b_1K_I)s^{m-1} \end{aligned} \tag{12}$$

$$
\begin{aligned}
&\ldots\ldots\ldots\ldots\ldots\ldots\ldots\ldots\ldots\ldots \\
&+(b_m K_D + b_{m-1} K_P + b_{m-2} K_I) s^2 \\
&+(b_{m+1} K_D + b_m K_P + b_{m-1} K_I) s^1 \\
&+(b_{m+2} K_D + b_{m+1} K_P + b_m K_I) s^0
\end{aligned}
.
$$

In terms of inner product, the general polynomial form of the closed-loop numerator polynomial is given by

$$
N^{CL}(s) = \sum_{i=0}^{m_{cl}} \langle K^{pid}, \overline{b}_{k-1} \rangle s^{m_{cl}-i}, \tag{13}
$$

where $m_{cl} = m + m^{pid}$ and vector $\overline{b}_k$ of the polynomial of zeros of the closed-loop system is given by

$$
\overline{b}_k = [b_k \quad b_{k-1} \quad b_{k-2}]. \tag{14}
$$

In similar way Eq. (13), one obtains the closed-loop denominator polynomial is given by

$$
\begin{aligned}
D^{CL}(s) = {} & s^{n+n_{D-pid}} \\
& + \sum_{i=0}^{n+n_{D-pid}-1} (b_k K_D + b_{k-1} K_P + b_{k-2} K_I + a_{i+1}),
\end{aligned} \tag{15}
$$

where $n_{D-pid} = 0$ or 1. When $n_{D-pid} = 0$, the PID controller structure have the terms derivative and proportional. When $n_{D-pid} = 1$, the structure of the PID controller has an integrator term that increases the order of the system by 1, starting with the three terms: proportional, derivative and integrative [21]. In this work, when $n_{D-pid} = 0$, the proposal is to specify an additional a_i^s coefficient., to ensure that the PID controller has the three terms.

3.3.2 Characteristic polynomial

The general form of the closed-loop denominator polynomial is given by

$$
P^{CL}(s) = s^{n_{cl}} + \sum_{i=1}^{n_{cl}-1} \left(a_i + \langle K^{pid}, \overline{b}_{k-1} \rangle\right) s^{n_{cl}-i}, \tag{16}
$$

where $P^{CL}(s)$ is the general form of the characteristic polynomial of the closed-loop plant and n_{cl} is the order of referred polynomial.

the characteristic closed-loop polynomial for unit feedback ($H(s) = 1$) is obtained in a similar way and given by

$$
P_p^{CL}(s) = 1 + C_{K^{pid}}(s) G(s). \tag{17}
$$

$P_p^{CL}(s)$ is the characteristic polynomial of the closed loop. The representation of the problem in the form of an internal product that relates the coefficients of the zero polynomial with the coefficients of the closed plant dynamics is the basis for obtaining the characteristic closed-loop polynomial.

3.4 Proposed method

The problem is formulated based on the propagation matrix generated from the dot product between the terms of the earnings vector K^{pid} with the coefficients of the numerator b_k associated with the coefficients of the denominator a_i of the TF of plant. The propagation matrix product associated with the TF numerator coefficients of the plant, give rise to a new characteristic polynomial based on new specified operating points, which are imposed by new zeros and new poles.

3.4.1 Propagation matrix of PID design

The design is based on the propagation matrix, allowing the designer to specify new points of operation that improve the performance of the controller acting on the plant dynamics, where changes in the order and coefficients of the characteristic polynomial can be observed through the internal product of the zero and gain coefficients of PID controllers.

The problem is formulated based on the propagation matrix $\overline{B}$, which is a consequence of the interaction between the parameters of the gain vector K^{pid} (K_D, K_P, K_I) with the coefficients of the plant TF numerator, this matrix is represented by

$$\overline{B} = \begin{bmatrix} b_0 & 0 & 0 \\ b_1 & b_0 & 0 \\ b_2 & b_1 & b_0 \\ b_3 & b_2 & b_1 \\ \cdots & \cdots & \cdots \\ b_m & b_{m-1} & b_{m-2} \\ 0 & b_m & b_{m-1} \\ 0 & 0 & b_m \end{bmatrix}. \tag{18}$$

One case notice in [20] that the diagonals are not repeated, and they vary according to the order n of the system's characteristic polynomial. The propagation of the gains is weighted by the coefficients of the numerator polynomial. The closed-loop TF of the plant is given in terms of the product of the gains K_D, K_P, and K_I with the coefficients b_k, b_{k-1}, and b_{k-2} of the closed-loop TF.

The law of formation of the propagation Matrix (18) is ruled by $m + 2$ rows and 3 columns. The rows represent the order of the system, starting with the propagation in the dynamics of order $s^{n_{cl}}$ and ending in the dynamics of order s^0 zero. The columns represent the gains of the controller in the poles and zeros of the plant dynamics.

3.4.2 Proposed characteristic polynomial

The proposed characteristic polynomial based on propagation matrix of PID controller gains idea is presented. From the system of equations that represents the actions of the PID controller in the plant dynamics, the formulation of the adjustment problem is established from the perspective of the inner product of the gains and the coefficients of the polynomials of the zeros of the closed-loop TF. In the case of the characteristic polynomial, the inner product is added to its coefficients. This way, the mechanism of gain adjustment is represented for allocations of zeros or poles.

From Eq. (17), the equation system that has an unknown vector K^{pid} and design specifications a_i^s, $i = 1, 2, ..., n+1$ is assembled. In scalar form, this system of equations is represented by

$$a_i + \langle K^{pid}, \overline{b}_k \rangle = a_i^s \Rightarrow \langle K^{pid}, \overline{b}_k \rangle = a_i^s - a_i \Rightarrow \langle K^{pid}, \overline{b}_k \rangle = a_i^e. \tag{19}$$

where $\overline{b}_k$ vector is assembled with the rows of the $\overline{B}$ matrix.

Expanding the scalar representation of Eq. (19), the system of equations to be solved is given by

$$\langle K^{pid}, \overline{b}_k \rangle = a_i^e \Rightarrow \begin{cases} K_D b_0 & + & K_P 0 & + & K_I 0 & = & a_1^e \\ K_D b_1 & + & K_P b_0 & + & K_I 0 & = & a_2^e \\ K_D b_2 & + & K_P b_1 & + & K_I b_0 & = & a_3^e \\ K_D b_m & + & K_P b_{m-1} & + & K_I b_{m-2} & = & a_4^e \\ K_D b_0 & + & K_P b_m & + & K_I b_{m-1} & = & a_5^e \\ \vdots & + & \vdots & + & \vdots & = & \vdots \\ K_D 0 & + & K_P 0 & + & K_I b_0 & = & a_n^e \end{cases} \tag{20}$$

The formulation of the problem presented in Eq. (19) and expanded in Eq. (20) is the starting point for the development of forms of parametric variation problems of TFs, as well as, for the establishment of operational points.

To determine the numerical values of the parameters K^{pid}, the following rules are presented: rule-1) the $\overline{B}$ matrix is assembled via Eq. (18), where b_k is the coefficients of TF numerator polynomial; rule-2) the new a_i^s parameters of the characteristic polynomial are specified; rule-3) the dot product of the parameters of the gain vector K^{pid} is made with the rows of the matrix $\overline{B}$, associated with the original a_i parameters of the characteristic polynomial and with the specified a_i^s parameters; rule-4) the system of equations given in Eq. (20) and rule-5 the system of equations given in rule-4 (Eq. (20)) is solved to determine the numerical values of the parameters of the gains vector K^{pid}.

4. Experiments

The experimental results are evaluated in three plants with mathematical models in terms of TF obtained with real data, being: Plant I of second order, with a zero; Third order plan II, with two zeros and fourth order plant III, with three zeros.

4.1 Plant I

Plant I, is a car dumper, which is used to unload solids in bulk, this equipment has the capacity to move up to 4,000 tons per hour (t/h). The general mathematical model of Plant I in TF is given by

$$G_{P_I}^G(s) = \frac{b_0}{s^2 + a_1 s + a_2}, \tag{21}$$

where, $G_{P_I}^G(s)$ is the TF of Plant I, it is a second order plant with zero at infinity.

The product of the TF numerator of Plant I given in Eq. (21) associated with the TF numerator of the controller given in Eq. (2) is given by

$$\left[C_{P_I}^{K^{pid}}(s)G_{P_I}^{G}(s)\right]_N = \left(K_D s^2 + K_P s + K_I\right)b_0 = K_D b_0 s^2 + K_P b_0 s + K_I b_0. \qquad (22)$$

The product of the TF denominator of Plant I given in Eq. (21) associated with the TF denominator of the controller given in Eq. (2) is given by

$$\left[C_{P_I}^{K^{pid}}(s)G_{P_I}(s)\right]_D = s\left(s^2 + a_1 s^2 + a_2 s\right) = s^3 + a_1 s^2 + a_2 s. \qquad (23)$$

The characteristic polynomial of Plant I is given by

$$P_I(s) = s^3 + (a_1 + K_D b_0)s^2 + (a_2 + K_P b_0)s + K_I b_0. \qquad (24)$$

System equations of Plant I related to Eq. (19) in the form $Ax = b$ is given by

$$\left\langle K^{pid}, \overline{b}_i \right\rangle = a_i^e \Rightarrow \begin{cases} a_1 + K_D b_0 = a_1^s; \\ a_2 + K_D b_0 = a_2^s; \\ a_3 + K_D b_0 = a_3^s; \end{cases}$$
$$\Rightarrow \begin{cases} K_D b_0 = a_1^s - a_1; \\ K_P b_0 = a_2^s - a_2; \\ K_I b_0 = a_3^s - a_3; \end{cases} \qquad (25)$$
$$\Rightarrow \begin{cases} 1)\ K_D b_0 = a_1^e; \\ 2)\ K_P b_0 = a_2^e; \\ 3)\ K_I b_0 = a_3^e. \end{cases}$$

Placing the systems of equations given in (25) in matrix form, we have

$$\left\langle K^{pid}, \overline{B} \right\rangle = a_i^e \Rightarrow \begin{bmatrix} b_0 & 0 & 0 \\ 0 & b_0 & 0 \\ 0 & 0 & b_0 \end{bmatrix} \times \begin{bmatrix} K_D \\ K_P \\ K_I \end{bmatrix} = \begin{bmatrix} a_1^e \\ a_2^e \\ a_3^e \end{bmatrix}. \qquad (26)$$

The transfer function of Plant I related to Eq. (21) is given by

$$G_{P_I}(s) = \frac{0.438}{1s^2 + 0.0861s + 0.0421s}. \qquad (27)$$

The a_i coefficients of the transfer function of the Plant - I related to Eq. (27) are given by

$$a_i \Rightarrow \begin{cases} a_1 = 0.0861; \\ a_2 = 0.0421; \\ a_3 = 0. \end{cases} \qquad (28)$$

The specified coefficients a_i^s of Plant I are given by

$$a_i^s \Rightarrow \begin{cases} a_1^s = 0.8604; \\ a_2^s = 0.421; \\ a_3^s = 0.0641. \end{cases} \qquad (29)$$

The error calculation a_i^e for Plant I associated with the coefficients a_i^s given in (29) and with the coefficients a_i given in (28) is given by

$$a_i^e \Rightarrow \begin{cases} a_1^s - a_1 = a_1^e = 0.8604 - 0.0861 = 0.7743; \\ a_2^s - a_2 = a_2^e = 0.421 - 0.0421 = 0.3789; \\ a_3^s - a_3 = a_3^e = 0.0641 - 0 = 0.0641. \end{cases} \tag{30}$$

The calculation of the K^{pid} gains vector of Plant I, is done by replacing the numerical values of Eq. (27) in the system of equations given in (25) and (26), is given by.

$$\begin{bmatrix} 0.438 & 0 & 0 \\ 0 & 0.438 & 0 \\ 0 & 0 & 0.438 \end{bmatrix} \times \begin{bmatrix} K_D \\ K_P \\ K_I \end{bmatrix} = \begin{bmatrix} 0.7743 \\ 0.3789 \\ 0.0641 \end{bmatrix}. \tag{31}$$

The Plant I given in Eq. (21) related to Eq. (21) has only the coefficient b_0, with that, the system of equations generated, related to the system of equations given in (25), has three equations and three unknowns.

$$K^{pid} \Rightarrow \begin{cases} i)\ 0.438K_D = 0.7749 \Rightarrow K_D = 0.7749/0.431 \Rightarrow K_D = 1.78; \\ ii)\ 0.438K_P = 0.3789 \Rightarrow K_P = 0.3789/0.431 \Rightarrow K_P = 0.87; \\ iii)\ 0.438K_I = 0.0641 \Rightarrow K_I = 0.0641/0.438 \Rightarrow K_I = 0.15. \end{cases} \tag{32}$$

Solving the system of equations given in (32), you can start with any of the equations to find the numerical values of K_D, K_P and K_I, since they are independent. With the numerical values of the earnings K_D, K_P and K_I, it replaces in the simulator developed in the MATLAB/SIMULINK software to monitor the performance of the proposed method.

Figure 2 shows the performance of the PID-Specified controller, which has the transfer function parameters specified by the designer and the $K_{Specified}^{pid}$ gain vector determined by the internal product of the vector of gains with the propagation matrix in purchase with the controller with the gains determined by the second method of ZN.

4.2 Plant II

Plant II, is a solid bulk reclaimer, which is used to recover bulk for ship loading, this equipment has the capacity to move up to 8,000 tons per hour (t/h).

$$G_{P_{II}}(s) = \frac{b_1 s + b_0}{s^3 + a_1 s^2 + a_2 s + a_3}, \tag{33}$$

where $G_{P_{II}}^G(s)$ is the TF of Plant II, it is a third order plant with zero at infinity.

The product of the TF numerator of Plant II given in Eq. (33) associated with the TF numerator of the controller given in Eq. (2) is given by

$$\begin{aligned} \left[C_{P_{II}}^{pid}(s)G_{P_{II}}(s)\right]_N &= \left(K_D s^2 + K_P s + K_I\right)(b_1 + b_0) \\ &= K_D b_1 s^3 + (K_D b_0 + K_I b_1)s^2 + K_I b_0. \end{aligned} \tag{34}$$

The product of the TF denominator of Plant II given in Eq. (33) associated with the TF denominator of the controller given in Eq. (2) is given by

$$\left[C_{P_{II}}^{pid}(s)G_{P_{II}}(s)\right]_D = s\left(s^3 + a_1s^2 + a_2s + a_3\right) = s^4 + a_1s^3 + a_2s^2 + a_3s. \quad (35)$$

The characteristic polynomial of Plant II is given by

$$\begin{aligned} P_{P_{II}}(s) = s^4 &+ (a_1 + K_D b_1)s^3 \\ &+ (a_2 + K_D b_0 + K_P b_1)s^2 \\ &+ (a_3 + (K_P b_0 + K_I b_1)s \\ &+ K_I b_0. \end{aligned} \quad (36)$$

System equations of Plant II related to Eq. (19) in the form $Ax = b$ is given by

$$\langle K^{pid}, \bar{b}_i \rangle = a_i^e \Rightarrow \begin{cases} a_1 + K_D b_1 = a_1^s; \\ a_2 + K_D b_0 + K_P b_1 = a_2^s; \\ a_3 + K_P b_0 + K_I b_1 = a_3^s; \\ a_4 + K_I b_0 = a_4^s; \end{cases} \Rightarrow \begin{cases} K_D b_1 = a_1^s - a_1; \\ K_D b_0 + K_P b_1 = a_2^s - a_2; \\ K_P b_0 + K_I b_1 = a_3^s - a_3; \\ K_I b_0 = a_4^s - a_4; \end{cases} \Rightarrow \begin{cases} 1)\ K_D b_1 = a_1^e; \\ 2)\ K_D b_0 + K_P b_1 = a_2^e; \\ 3)\ K_P b_0 + K_I b_1 = a_3^e; \\ 4)\ K_I b_0 = a_4^e. \end{cases} \quad (37)$$

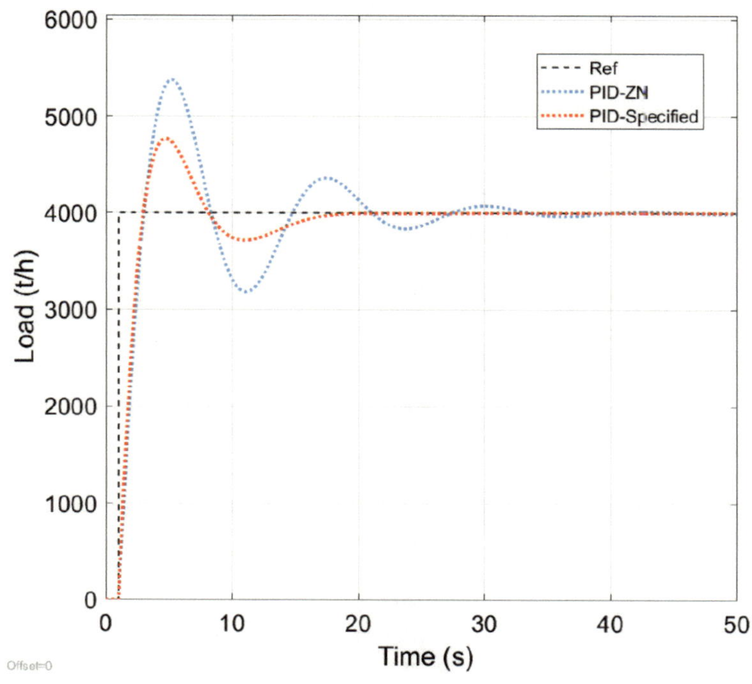

Figure 2.
Plant I - PID-ZN x PID-specified.

Placing the systems of equations given in (37) in the matrix form, we have

$$\langle K^{pid}, \overline{B} \rangle = a_i^e \Rightarrow \begin{bmatrix} b_0 & 0 & 0 \\ b_1 & b_0 & 0 \\ 0 & b_1 & b_0 \\ 0 & 0 & b_1 \end{bmatrix} \times \begin{bmatrix} K_D \\ K_P \\ K_I \end{bmatrix} = \begin{bmatrix} a_1^e \\ a_2^e \\ a_3^e \\ a_4^e \end{bmatrix}. \quad (38)$$

The transfer function of Plant II related to Eq. (33) is given by

$$G_{P_{II}}(s) = \frac{0.1812s + 0.087}{s^3 + 0.3853s^2 + 0.117s + 0.01567}. \quad (39)$$

The a_i coefficients of the Plant TF - II related to Eq. (39) are given by

$$a_i \Rightarrow \begin{cases} a_1 = 0.3853; \\ a_2 = 0.117; \\ a_3 = 0.01567; \\ a_4 = 0. \end{cases} \quad (40)$$

The specified coefficients a_i^s of Plant II are given by

$$a_{4i}^{e\,s} \Rightarrow \begin{cases} a_1^s = 1.7133; \\ a_2^s = 0.8542; \\ a_3^s = 0.2670; \\ a_4^s = 0.0478. \end{cases} \quad (41)$$

The error calculation a_i^e for Plant II associated with the coefficients a_i^s given in (41) and with the coefficients a_i given in (40) is given by

$$a_{4i}^{e\,e} \Rightarrow \begin{cases} a_1^e = a_1^s - a_1 = 1.7133 - 0.3853 = 1.328; \\ \mathrm{a}_2^e = a_2^s - a_2 = 0.7372 - 0.117 = 0.6202; \\ a_3^e = a_3^s - a_3 = 0.2514 - 0.01567 = 0.2358; \\ a_4^e = a_4^s - a_4 = 0.0478 - 0 = 0.0478. \end{cases} \quad (42)$$

The calculation of the K^{pid} gains vector is performed by replacing the numerical values of Eq. (39) in the system of equations given in (37) and (38).

$$\begin{bmatrix} 0.087 & 0 & 0 \\ 0.1812 & 0.087 & 0 \\ 0 & 0.1812 & 0.087 \\ 0 & 0 & 0.1812 \end{bmatrix} \times \begin{bmatrix} K_D \\ K_P \\ K_I \end{bmatrix} = \begin{bmatrix} 1.3280 \\ 0.9538 \\ 0.0996 \\ 0.0478 \end{bmatrix}. \quad (43)$$

The plant II given in Eq. (39) related to Eq. (33) has the b_0 coefficients and b_1, with this, the system of equations generated, for the system of equations related to the system of equations given in (37), has four equations and three unknowns.

$$K^{pid} \Rightarrow \begin{cases} i)\, K_D b_1 = a_1^e \Rightarrow K_D = 1.328/0.1812 \Rightarrow K_D = 7.329; \\ ii)\, K_D b_0 + K_P b_1 = a_2^e \Rightarrow K_P = (0.9538 - 0.6376)/0.1812 \Rightarrow K_P = 1.745; \\ iii)\, \mathrm{K}_P b_0 + K_I b_1 = a_3^e \Rightarrow K_I = (2514 - 0.1518)/0.1812 \Rightarrow K_I = 0.549; \\ iv)\, K_I b_0 = a_4^e \Rightarrow K_I = 0.0478/0.087 \Rightarrow K_I = 0.549. \end{cases} \quad (44)$$

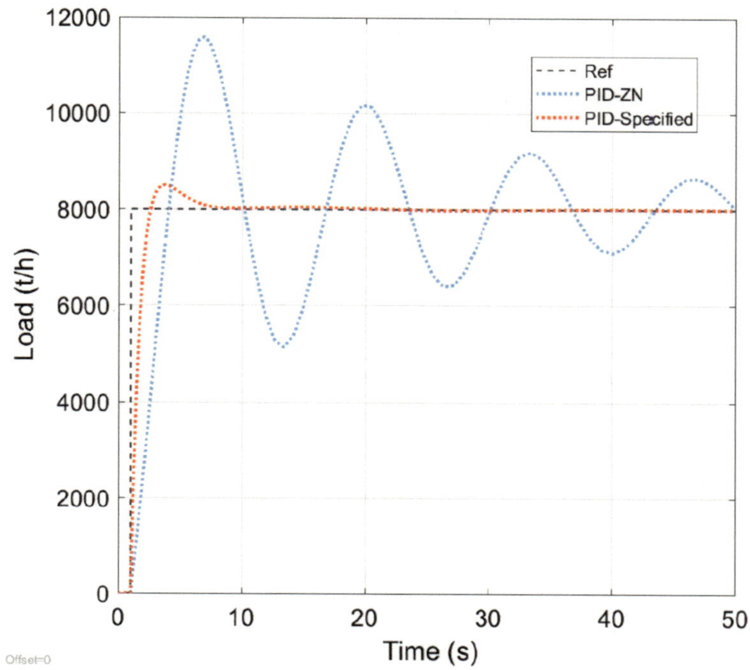

Figure 3.
Plant II - PID-ZN x PID-specified.

Solving the system of equations given in (44), first, solve Equation i) to find the numerical value of K_D. Then, replace the numerical value of K_D in Equation ii) and find the numerical value of K_P. To find the numerical value of K_I, solve equation iv) or replace the values of K_P in equation iii). With the numerical values of the gains K_D, K_P and K_I, it replaces in the simulator developed in the MATLAB/SIMULINK software to monitor the performance of the proposed method.

Figure 3 shows the performance of the PID-Specified controller, which has the transfer function parameters specified by the designer and the $K^{pid}_{Specified}$ gain vector determined by the internal product of the vector of gains with the propagation matrix in purchase with the controller with the gains determined by the second method of ZN.

4.3 Plant III

Plant III, is a car dumper with two feeders, which is used to unload solids in bulk, this equipment has the capacity to move up to 8,000 tons per hour (t/h). The general mathematical model of Plant III in TF is given by

$$G^G_{P_{III}}(s) = \frac{b_2 s^2 + b_1 s + b_0}{s^4 + a_1 s^3 + a_2 s^2 + a_3 s + a_4}. \tag{45}$$

where, $G^G_{P_{III}}(s)$ is the TF of Plant III, it is a fourth order plant with zero at infinity.

The product of the TF numerator of Plant III given in Eq. (45) associated with the TF numerator of the controller given in Eq. (2) is given by

$$\begin{aligned}\left[C^{pid}_{P_{III}}(s)G_{P_{III}}(s)\right]_N &= \left(K_Ds^2+K_Ps+K_I\right)\left(b_2s^2+b_1s+b_0\right)\\ &= K_Db_2s^4+(K_Db_1+K_Pb_2)s^3\\ &\quad +(K_Db_0+K_Pb_1+K_I)s^2\\ &\quad +(K_Pb_0+K_Ib_1)s\\ &\quad +K_Ib_0.\end{aligned} \tag{46}$$

The product of the TF denominator of Plant III given in Eq. (45) associated with the TF numerator of the controller given in Eq. (2) is given by

$$\begin{aligned}\left[C^{pid}_{P_{III}}(s)G_{P_{III}}(s)\right]_D &= s\left(s^4+a_1s^3+a_2s^2+a_3s+a_4\right)\\ &= s^5+a_1s^4+a_2s^3+a_3s^2+a_4s.\end{aligned} \tag{47}$$

The characteristic polynomial of Plant III is given by

$$\begin{aligned}P_{P_{III}}(s) = s^5 &+ a_1 + K_Db_2s^4\\ &+a_2+(K_Db_1+K_Pb_2)s^3\\ &+a_3+(K_Db_0+K_Pb_1+K_Ib_2)s^2\\ &+a_4+(K_Pb_0+K_Ib_1)s\\ &+K_Ib_0.\end{aligned} \tag{48}$$

System equations da Planta III related to Eq. (19) in the form $Ax=b$ is given by

$$\begin{aligned}\left\langle K^{pid},\overline{b}_i\right\rangle = a_i^e \Rightarrow &\begin{cases} a_1+K_Db_2=a_1^s;\\ a_2+K_Db_1+K_Pb_2=a_2^s;\\ a_3+K_Db_0+K_Pb_1+K_Ib_2=a_3^s;\\ a_4+K_Pb_0+K_Ib_1=a_4^s;\\ a_5+K_Ib_0=a_5^s.\end{cases}\\ \Rightarrow &\begin{cases} K_Db_2=a_1^s-a_1;\\ K_Db_1+K_Pb_2=a_2^s-a_2;\\ K_Db_0+K_Pb_1+K_Ib_2=a_3^s-a_3;\\ K_Pb_0+K_Ib_1=a_4^s-a_4;\\ K_Ib_0=a_5^s-a_5.\end{cases}\\ \Rightarrow &\begin{cases} 1)K_Db_2=a_1^e;\\ 2)K_Db_1+K_Pb_2=a_2^e;\\ 3)K_Db_0+K_Pb_1+K_Ib_2=a_3^e;\\ 4)K_Pb_0+K_Ib_1=a_4^e;\\ 5)K_Ib_0=a_5^e.\end{cases}\end{aligned} \tag{49}$$

Placing the systems of equations given in (49) in the matrix form, we have

$$\left\langle K^{pid},\overline{B}\right\rangle = a_i^e \Rightarrow \begin{bmatrix} b_0 & 0 & 0\\ b_1 & b_0 & 0\\ b_2 & b_1 & b_0\\ 0 & b_2 & b_1\\ 0 & 0 & b_2\end{bmatrix} \times \begin{bmatrix} K_D\\ K_P\\ K_I\end{bmatrix} = \begin{bmatrix} a_1^e\\ a_2^e\\ a_3^e\\ a_4^e\\ a_5^e\end{bmatrix}. \tag{50}$$

The transfer function of Plant III related to Eq. (45) is given by

$$G_{P_{III}}(s) = \frac{0.959s^2 + 0.1698s + 0.1593}{s^4 + 0.1767s^3 + 0.3463s^2 + 0.029s + 0.02331}. \tag{51}$$

The a_i coefficients of the Plant TF - III related to Eq. (51) are given by

$$a_i \Rightarrow \begin{cases} a_1 = 0.1767; \\ a_2 = 0.3463; \\ a_3 = 0.029; \\ a_4 = 0.2331; \\ a_5 = 0. \end{cases} \tag{52}$$

The specified coefficients a_i^s of Plant III are given by

$$a_i^s \Rightarrow \begin{cases} a_1^s = 1.8358; \\ a_2^s = 8.619; \\ a_3^s = 8.9947; \\ a_4^s = 1.4277; \\ a_5^s = 1,3254. \end{cases} \tag{53}$$

The error calculation a_i^e for Plant III associated with the coefficients a_i^s given in (53) and with the coefficients a_i given in (52) is given by

$$a_i^e \Rightarrow \begin{cases} a_1^e = a_1^s - a_1 = 1.8358 - 0.1767 = 1.659; \\ a_2^e = a_2^s - a_2 = 8.619 - 0.3463 = 4.9066; \\ a_3^e = a_3^s - a_3 = 8.9947 - 0.029 = 8.9657; \\ a_4^e = a_4^s - a_4 = 1.4277 - 0.0233 = 1.4044; \\ a_5^e = a_5^s - a_5 = 1.3254 - 0 = 1.3254. \end{cases} \tag{54}$$

The calculation of the gain vector K^{pid} is done by replacing the numerical values of Eq. (51) in the system of equations given in (49) and (50).

$$\begin{bmatrix} 0.1593 & 0 & 0 \\ 0.1698 & 0.1593 & 0 \\ 0.959 & 0.1698 & 0.1593 \\ 0 & 0.959 & 0.1698 \\ 0 & 0 & 0.959 \end{bmatrix} \times \begin{bmatrix} K_D \\ K_P \\ K_I \end{bmatrix} = \begin{bmatrix} 1.6591 \\ 4.9066 \\ 9.0712 \\ 2.7381 \\ 1.3254 \end{bmatrix}. \tag{55}$$

The Plant III given in Eq. (51) related to Eq. (45), has the coefficients b_0, b_1 and b_2, with that, the generated system of equations, for the system of equations related to the system of equations given in (49), has five equations and three unknowns.

$$K^{pid} \Rightarrow \begin{cases} i)\; K_D b_2 = a_1^e \Rightarrow K_D = 1.6591/959 \Rightarrow K_D = 1.73; \\ ii)\; K_D b_1 + K_P b_2 = a_2^e \Rightarrow K_P = 4.6128/0.959 \Rightarrow K_P = 4.81; \\ iii)\; K_D b_0 + K_P b_1 + K_I b_2 = a_3^e \Rightarrow K_I = 7.978/0.959 \Rightarrow K_I = 8.32; \\ iv)\; K_P b_0 + K_I b_1 = a_4^e \Rightarrow K_I = 1.4127/0.1698 \Rightarrow K_I = 8.32; \\ v)\; K_I b_0 = a_5^e \Rightarrow K_I = 1.3254/0.1593 \Rightarrow K_I = 8.32. \end{cases} \tag{56}$$

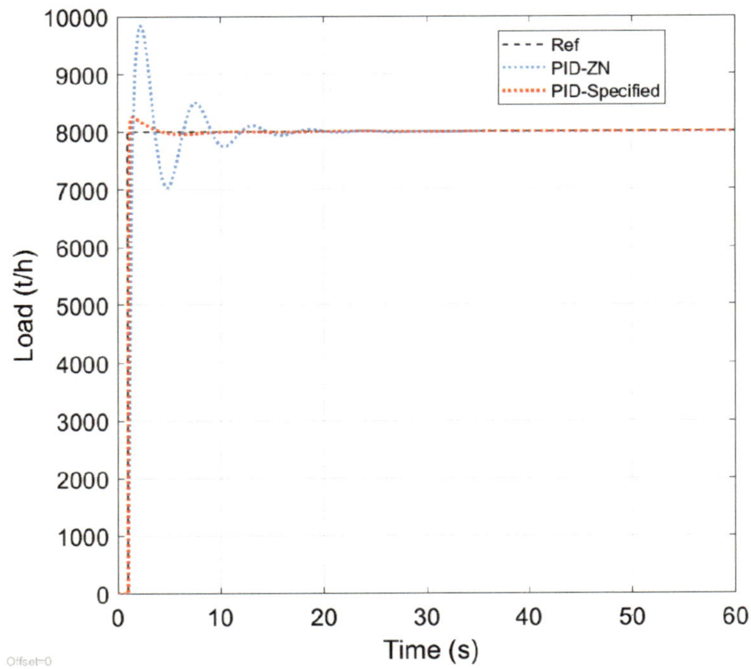

Figure 4.
Plant III - PID-ZN x PID-specified.

Solving the system of equations given in (56), first, solve Equation i) to find the numerical value of K_D. Then, replace the numerical value of K_D in Equation ii) and find the numerical value of K_P. To find the numerical value of K_I, solve Equation v) or replace the values of K_D and K_P in Equation iii) or you can substitute the value of K_P in Equation iv). With the numerical values of the gains K_D, K_P and K_I, it replaces in the simulator developed in the MATLAB/SIMULINK software to monitor the performance of the proposed method.

Figure 4 shows the performance of the PID-Specified controller, which has the transfer function parameters specified by the designer and the $K^{pid}_{Specified}$ gain vector determined by the internal product of the vector of gains with the propagation matrix in purchase with the controller with the gains determined by the second method of ZN.

5. Conclusions

The study presented a methodology for adjusting the gains of a PID controller in terms of the internal product of the gains and the propagation matrix. In addition, the relevance of the matrix was shown, which enabled impact assessment of the PID actions associated with the plant parameters. The proposed methodology complied with the project specifications and ensured high controller efficiency without suppressing the PID terms caused by the adjustments. The three PID controller terms were adjusted. Therefore, this methodology can be considered as an alternative to conventional methods for the computation of K_D, K_P, and K_I gains parameters in practical applications and highly complex control plants.

Acknowledgements

The authors would like to thank PPGEE of the UFMA for the resources to develop this work. We are especially grateful to FAPEMA for research incentive and infrastructure. We acknowledge the Department of Computer Engineering of the UEMA for making this research possible. Finally, we also acknowledge CAPES and CNPq for promoting and supporting the advanced studies that contributed to this work.

Author details

José Pinheiro de Moura[1,2*†] and João Viana da Fonseca Neto[2†]

1 State University of Maranhão, Brazil

2 Federal University of Maranhão, Brazil

*Address all correspondence to: josepinheiro@professor.uema.br

† These authors contributed equally.

References

[1] Sutikno Juwari Purwo and Hidayah Nur and Handogo Renanto. Maximum Peak-Gain Margin (Mp-GM) Tuning Method for Two Degree of Freedom PID Controller. ID Control for Industrial Processes. BoD–Books on Demand, 2018.

[2] Bucz Štefan and Kozáková Alena. Advanced methods of PID controller tuning for specified performance. PID Control for Industrial Processes. IntechOpen London, 2018.

[3] Jeng Jyh-Cheng. Data-based tuning of PID controllers: a combined model-reference and VRFT method. PID Control for Industrial Processes. BoD–Books on Demand, 2018.

[4] Moura J P, Fomseca Neto J V, Rêgo P H M. A Neuro-Fuzzy Model for Online Optimal Tuning of PID Controllers in Industrial Systems Applications to the Mining Sector. IEEE - Transactions on Fuzzy Systems, 2019.

[5] Moura J P, Fomseca Neto, Ferreira E F M, Araujo Filho E. M. On the Design and Analysis of Structured-ANN for Online PID-Tuning to Bulk Resumption Process in Ore Mining System. Elsevier - Neurocomputing, 2020.

[6] Chen, Chi-Tsong. Linear System Theory and Design. Inc. 3rd. Oxford University Press, New York, NY, USA, 1998

[7] Åström K J, Wittenmark B. Computer-Controlled Systems: Theory and Design, Third Edition, isbn: 9780486284040, series: Dover Books on Electrical Engineering. Dover Publications, 2013.

[8] Vilanova R, Visioli A. PID Control in the Third Millennium: Lessons Learned and New Approaches, isbn: 9781447124252, series: Advances in Industrial Control. Springer London, 2012.

[9] Tavakoli, Ali Reza and Seifi, Ali Reza. Adaptive self-tuning PID fuzzy sliding mode control for mitigating power system oscillations. Neurocomputing, volume 218, pages 146–153. Elsevier, 2016.

[10] Liu P, Gu, Haibo, Kang, Yu and Lü, Jinhu. Global synchronization under PI/PD controllers in general complex networks with time-delay. Neurocomputing. Elsevier, 2019.

[11] Aidan O'Dwyer. Handbook Of Pi And Pid Controller Tuning Rules, isbn: 1848162421,9781848162426. Imperial College Press, 2009.

[12] Alfaro, V.M. and Vilanova, R. Model-Reference Robust Tuning of PID Controllers, isbn: 9783319282138, lccn: 2016935592, series: Advances in Industrial Control. Springer International Publishing, 2016.

[13] S. Bennett. A brief history of automatic control. doi 10.1109/37.506394, issn: 1066-033X. IEEE Control Systems, volume 16, number 3, pages 17–25, 1996.

[14] Lee, Jay and Kao, Hung-An and Yang, Shanhu. Service innovation and smart analytics for industry 4.0 and big data environment. Procedia Cirp, volume 16, pages 3–8. Elsevier, 2014.

[15] Lasi, Heiner and Fettke, Peter and Kemper, Hans-Georg and Feld, Thomas and Hoffmann, Michael. Industry 4.0. Business & Information Systems Engineering, volume 6, number 4, pages 239–242, Springer, 2014.

[16] Marwedel, P. Embedded System Design: Embedded Systems Foundations of Cyber-Physical Systems, and the Internet of Things, isbn: 9783319560458, series: Embedded Systems. Springer International Publishing, 2017.

[17] Kopetz, H. Real-Time Systems: Design Principles for Distributed Embedded Applications, isbn: 9781441982377, series: Real-Time Systems Series. Springer US, 2011.

[18] Fadali, M.S. and Visioli, A. Digital Control Engineering: Analysis and Design, isbn: 9780123983244, lccn: 2012021488. lsevier Science, 2012.

[19] Quddious Abdul and Antoniades Marco A and Vryonides Photos and Nikolaou Sym. Voltage-Doubler RF-to-DC Rectifiers for Ambient RF Energy Harvesting and Wireless Power Transfer Systems. Wireless Energy Transfer Technology. IntechOpen, 2019.

[20] Moura, José Pinheiro, Fonseca Neto, João Viana and Rêgo, Patrícia Helena M. Models for Optimal Online Tuning Based on Computational Intelligence of PID Controllers Applied to Operational Processes of Bulk Reclaimers, issn: 2195–3899, doi:10.1007/s40313-018-00438-7. Journal of Control, Automation and Electrical Systems. Springer, 2019.

[21] Moura, José Pinheiro and Neto, João Viana Fonseca and Ferreira, Ernesto Franklin Marçal and Araujo Filho, Evandro Martins. On the Design and Analysis of Structured-ANN for Online PID-Tuning to Bulk Resumption Process in Ore Mining System. Neurocomputing. Elsevier, 2020.

Chapter 5

Enhanced Nonlinear PID Controller for Positioning Control of Maglev System

Shin-Horng Chong, Roong-Soon Allan Chan and Norhaslinda Hasim

Abstract

Magnetic levitation (maglev) is a way of using electromagnetic fields to levitate objects without any noise or the need for petrol or air. Due to its highly nonlinear and unstable behavior, numerous control solutions have been proposed to overcome it. However, most of them still acquire precise dynamic model parameters, or deep understanding of control theory. To account the complexity in the design procedure, a practical controller consists of classical and modern control approaches are proposed. This chapter presents a practical controller for high positioning performance of a magnetic levitation system. Three strategies of the proposed controller where the PI-PD controller is to enhance transient response, the model-based feedforward control (FF) is incorporated with the PI-PD controller to enhance the overshoot reduction characteristic in attaining a better transient response, and lastly the disturbance compensator (K_z) is integrated as an additional feedback element to reduce the sensitivity function magnitude for robustness enhancement. The proposed controller - FF PI-PD + K_z has a simple and straightforward design procedure. The usefulness of the proposed controller is evaluated experimentally.

Keywords: maglev system, disturbance compensator, model-based feedforward control, PI-PD control, robustness

1. Introduction

Magnetic levitation (maglev) system produces an electromagnetic force as the electric current flow through the coils to support a levitated object. This indicates that the maglev system eliminates the mechanical contact and friction between the moving and stationary parts. Due to the advantages, the maglev system has been successfully and widely implemented for many high-speed motion applications such as the high-speed maglev passenger trains, magnetic bearing system, flywheel energy storage system and vibration isolation system [1]. However, the maglev system is open loop instability and inherent nonlinearities. In addition, it is a non-damping system which has fast response, yet sensitive to vibration. Therefore, it remains a challenging task to design a feedback controller for attaining a good positioning performance in the maglev system.

Although a lot of advanced control strategy has been proposed for controlling maglev system, the classical controllers such as proportional integral derivative

(PID) and lead–lag compensators still are regularly employed in the industrial applications due to their simple structure, straightforward design procedure and easy to implement. In the past, a lead compensator [2] and cascaded lead compensation scheme [3] were designed to stabilize the maglev system. However, the classical controllers can only perform well in limited operating range and failed to demonstrate a satisfactory robustness performance. Thus, many advanced controllers such as feedback linearization [4], sliding mode control [5], H_∞ control technique [6], disturbance observer control approach [7], adaptive robust backstepping method [8] and model predictive control [9–11] as well as intelligent controllers which include fuzzy logic control [12] and neural network [13] have been dedicated to procure the high positioning and robustness performances in the maglev system. Despite the good positioning and robustness performances of the advanced controllers, sufficient knowledge of control theory is strictly needed in the design procedure. Furthermore, the intelligent controllers consist of complex architecture require a high computational effort. Often, the intelligent controllers do not have any systematic design procedure. These drawbacks depict barrier to their practical use.

Due to the above-mentioned reason, researchers kept devoting their effort in enhancing control performance of classic control in maglev system. The problem associated with 1-DOF PID control was overcome with the proposed of modified PID control and/or 2-DOF PID control. In 2007, Leva and Bascetta in [14] have realized a model-based feedforward control to the PID controller to improve the tracking performance of a maglev system. Unfortunately, it still demonstrated huge spike occurrence when the ball started moving at the initial position. After few years, Ghosh et al. has proposed a 2-DOF PID controller to improve the system transient response with zero percentage of overshoot. However, the proposed controller suffered from long settling time which was around 2 s. Besides, its positioning accuracy was recorded poor due to the derivative action on the reference signal [15]. In order to solve the positioning accuracy, Allan et al. has introduced the 2-DOF Lead-plus-PI controller [16]. The experimental evidence reported the improvement in the positioning accuracy, yet to point-to-point motion performance was deteriorated as the levitation height was increased.

Thus, in this research, a proportional integral-proportional derivative control with feedforward and disturbance compensations (FF PI-PD + K_z) control approach is proposed to stabilize the maglev system and enhance the positioning performance as well as its robustness. The proposed controller consists of a PI-PD controller, a model-based feedforward control and a disturbance compensator. The PI-PD controller is designed by using the pole-placement method; the model-based feedforward control is constructed based on the system driving characteristic in open loop; the disturbance compensator is developed via the system current dynamics in closed loop. The derivative action of the PI-PD control amplified the measurement noise that affected the positioning accuracy. Hence, a low pass filter is featured with the PI-PD control to suppress the bad influence of the derivative action. Besides, a model-based feedforward control is incorporated with the PI-PD controller to further improve the following characteristic of the mechanism in attaining a better overshoot reduction characteristic. At the same time, the positioning time is greatly reduced. Lastly, a disturbance compensator is integrated as an additional feedback element for robustness enhancement via lowering the sensitivity function magnitude. The effectiveness of the FF PI-PD + K_z controller is validated experimentally through two types of motion control that are point-to-point and tracking motions. In this present paper, the robustness of the FF PI-PD + K_z controller is examined via applying an impulse disturbance and varying the mass. The positioning and robustness performances of the FF PI-PD + K_z controller are compared with the FF PI-PD and the full state feedback (FSF) controllers.

The remainder of this chapter is outlined as follow. In Section 2, the experimental setup and mathematical modeling of the maglev system are represented. Section 3 explains the control structure, design procedure and stability analysis of the FF PI-PD + K_z controller. In Section 4, the experimental results are discussed. Lastly, the conclusion is drawn in Section 5.

2. Experimental setup and dynamic modeling

This section presents the experimental setup and dynamic modeling of the maglev system.

2.1 Experimental setup

The single-axis maglev mechanism (Googoltech GML 2001), as shown in **Figure 1** is used as a testbed to clarify the usefulness of the proposed controller. The maglev system is only able to control object to move up and down. The control purpose is to keep the magnetic levitation ball stable in a given position or to make the ball track a desired trajectory.

The maglev mechanism consists of an electromagnet (number of windings, N_w = 1000 turns) to exert a tractive force across the air gap to levitate a steel ball (mass, M = 94 g). Besides, it is a voltage-controlled (control signal, u = 0–10 V) maglev mechanism, which is comprised of a power amplifier to actuate the electromagnet. The maximum electrical power consumption is around 16 W. The maximum levitation height of the maglev system is 15 mm. In the experiments, the initial position is set at 10.5 mm and the operating range is within ±2.5 mm. A laser position sensor (Panasonic laser distance sensor HG-C1050) with resolution of 1.83 μm is used to measure the levitation displacement. As experimentally examined, the resolution of the laser sensor output in open loop is recorded at 15 μm. To measure the controlled current of the mechanism, a hall effect current sensor

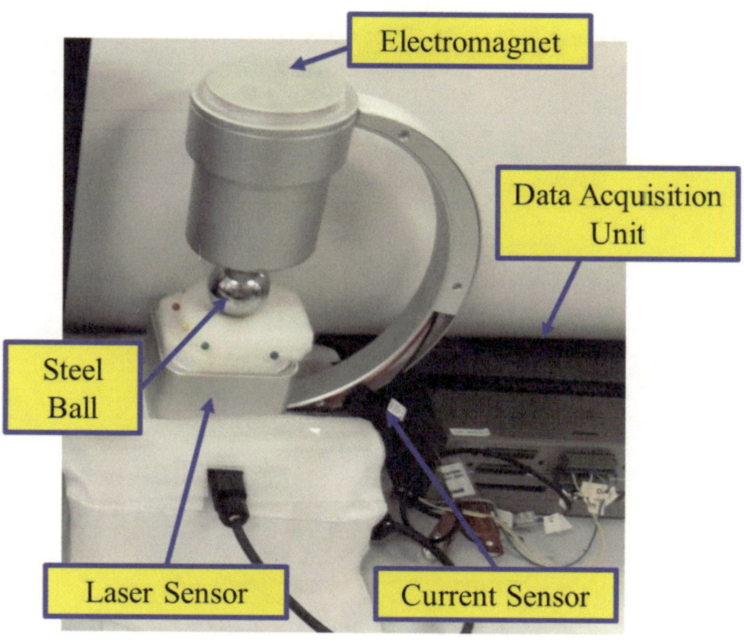

Figure 1.
Maglev system.

(Magnelab hall effect current sensor HCT-0010-005) with resolution of 0.38 mA is used. The controller is implemented at a sampling rate of 1 kHz.

2.2 Dynamic modeling

Figure 2 illustrates the principle diagram of maglev system. The dynamic equation of the maglev system is

$$M\frac{d^2x}{dt^2} = -F_m(i,x) + F_g \tag{1}$$

where M, F_{m}, F_g, i and x denote the steel ball mass, electromagnetic force, gravitational force, current and levitation height respectively.

The F_m (i, x) is in negative sign indicates that it always functioning in opposite direction against the gravitational force, F_g.

The electromagnetic force, $F_m(i, x)$ is described as

$$F_m(i,x) = K\frac{i^2}{x^2} \tag{2}$$

where K represents the electromagnetic constant.

The gravitational force, F_g is denoted as

$$F_g = Mg \tag{3}$$

where g represents the gravitational acceleration.

Substitute Eqs (2) and (3) into Eq. (1), the dynamic equation of the maglev system can be accordingly rewritten as

$$\frac{d^2x}{dt^2} = -\frac{K}{M}\frac{i^2}{x^2} + g \tag{4}$$

The Eq. (2) shows the inherent nonlinearities characteristic of the F_m (i, x) which can be linearized by using the Taylor Series approximation at the equilibrium position where

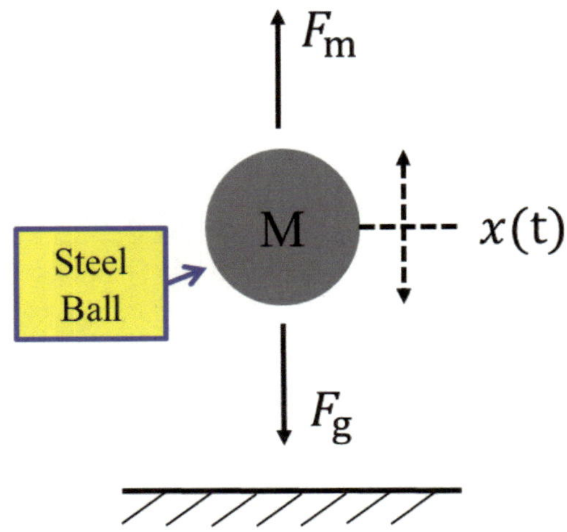

Figure 2.
The principal diagrams of the maglev system.

$$F_m(i,x) \approx F_m(i_o,x_o) + \frac{\partial F_m(i,x)}{\partial i} i(t) + \frac{\partial F_m(i,x)}{\partial x} x(t) \tag{5}$$

During levitating, the relationship between the $F_m\ (i, x)$ and the F_g is governed by

$$M_g = F_m(i_o,x_o) \tag{6}$$

where i_o and x_o denote the nominal current and nominal displacement, respectively.

Substitute Eqs (5) and (6) into Eq. (1), and undergoes Laplace transform on Eq. (1), the linear transfer function is

$$\frac{X(s)}{I(s)} = \frac{-\frac{K_c}{M}}{\left[s^2 - \frac{K_x}{M}\right]} \tag{7}$$

where K_c and K_x represent the current and position coefficient, while $X(\mathrm{s})$ and $I(\mathrm{s})$ denote the levitation height and current, respectively, where $K_c = 2Ki_o / x_o^2$ and $K_x = 2Ki_o^2/x_o^3$.

Rewrite Eq. (7) by involving power amplifier gain, K_a and sensor sensitivity gain, K_s, it becomes

$$G_p(s) = \frac{V_s(s)}{V_i(s)} = \frac{-\frac{K_s}{K_a}\frac{K_c}{M}}{\left[s^2 - \frac{K_x}{M}\right]} \tag{8}$$

The Eq. (8) is amplified to

$$G_p(s) = \frac{\beta}{s^2 - \gamma^2} \tag{9}$$

where $\beta = -K_sK_c/K_aM$ and $\gamma^2 = K_x/M$.

As shown in Eq. (9), it proves that the uncompensated system is unstable in open loop because it comprises of one pole located at the right half s-plane. Thus, a feedback control system is a vital need to stabilize the system. The system parameter values are shown in **Table 1.**

3. Proportional integral-proportional derivative control with feedforward and disturbance compensations (FF PI-PD + *Kz*) control system framework

This section is devoted to explaining the formulation of FF PI-PD + K_z control approach for the maglev system. Next, the control strategy of FF PI-PD + K_z controller is discussed and followed by the design procedure of the proposed control. Lastly, the stability of the proposed control is examined.

3.1 Control structure

The block diagram of FF PI-PD + K_z control system for positioning and robust control of 1-DOF maglev system is depicted in **Figure 3**. The feedback loop consists of PI-PD control and disturbance compensation scheme, whereas the feedforward loop contains a model-based feedforward control. The FF PI-PD + K_z control system

Symbol	Description, unit	Value
M	Steel ball mass, Kg	9.40×10^{-2}
x_o	Nominal displacement, m	1.00×10^{-2}
i_o	Nnominal current, A	3.94×10^{-1}
K	Electromagnetic constant, Nm^2/A^2	2.31×10^{-4}
K_a	Power amplifier gain, V/A	6.51
K_s	Sensor sensitivity, V/m	1.67×10^{2}
g	Gravitational acceleration, m/s^2	9.81

Table 1.
Model parameters.

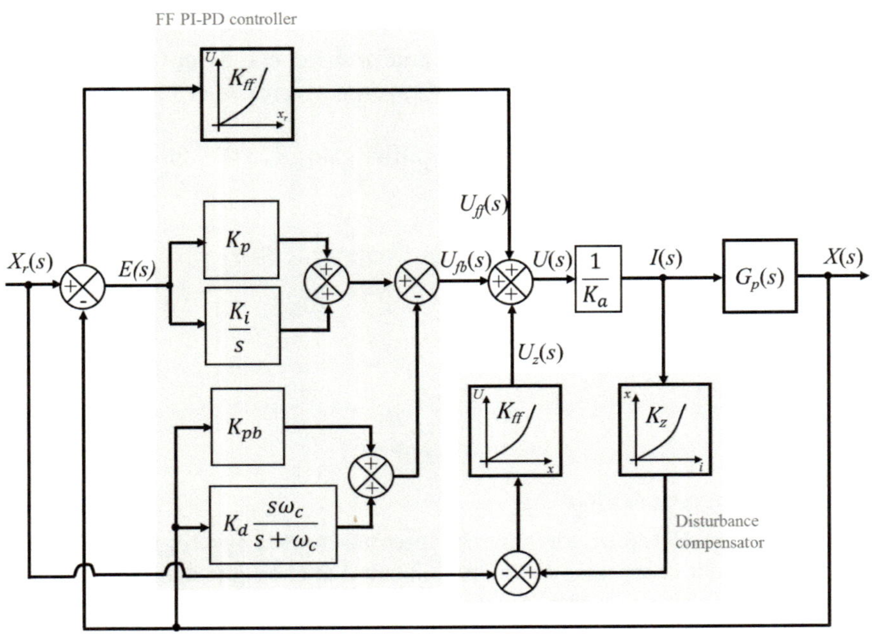

Figure 3.
Block diagram of the FF PI-PD + Kz control system.

is designed under the following considerations: (i) the PI-PD control is designed to improve the transient response of the conventional PID controller, (ii) the model-based feedforward control is integrated to obtain a better overshoot reduction characteristic and (iii) the disturbance compensation control is employed for robustness enhancement.

To design the FF PI-PD + K_z controller, the PI-PD control is designed at the first place. The PI-PD controller is proposed to improve the positioning performance of the conventional PID controller [17]. By moving the derivative action and some portion of the proportional gain to the feedback path, the resonance peak of conventional PID controller in the closed-loop frequency response can be reduced. Thus, it explains that the PI-PD controller demonstrates a better transient response than the conventional PID controller. Then, a low pass filter is adopted to improve the positioning accuracy by attenuating the amplification of the measurement noise. However, the PI-PD control transient response is unsatisfied because the overshoot remains high. To solve this problem, a model-based feedforward control is incorporated with the PI-PD control to enhance the system following characteristic. This improvement leads to a better overshoot reduction characteristic and shorten the positioning time. Lastly, a disturbance compensator is introduced for robustness enhancement of the FF PI-PD control. The proposed disturbance compensator estimates the disturbances and imparts an adequate voltage to compensate them. Overall, the proposed FF PI-PD + K_z control system provides the advantages which are: (i) a better overshoot reduction characteristic; and (ii) low sensitivity to external disturbance and parameter variation.

The control law of the FF PI-PD + Kz control system is

$$U(s) = \left(K_p + \frac{K_i}{s}\right)E(s) - \left(K_{pb} + K_d\frac{s\omega_c}{s+\omega_c}\right)X(s) + K_{ff}X_r(s) + K_{ff}[K_zI(s) - X_r(s)] \tag{10}$$

where K_p, K_i, K_{pb}, K_d, K_{ff}, K_z, ω_c, *and* X_r denote the proportional gain, integral gain, feedback proportional gain, derivative gain, linearized feedforward gain, linearized disturbance compensator gain, system cut-off frequency and reference input, respectively.

3.2 Design procedure

There are three (3) major parts in the design procedure of the FF PI-PD + Kz control system.

3.2.1 PI-PD controller

The PI-PD controller is a modified PID controller, where it consists of derivative gain, K_d and some portion of the proportional gain, K_{pb} at the feedback path. Both are evidenced to have an approximately similar closed loop characteristic equation, as proved in Eqs (11) and (12).

$$\delta_{PID}(s) = s^3 + \beta K_d s^2 + (K_p\beta - \gamma^2)s + K_i\beta \tag{11}$$

$$\delta_{PI-PD}(s) = s^3 + \beta K_d s^2 + \left[(K_p + K_{pb})\beta - \gamma^2\right]s + K_i\beta \tag{12}$$

The desired characteristic equation of a general second-order system is denoted as

$$\delta_{desired}(s) = (s + \alpha\zeta\omega_n)\left(s^2 + 2\zeta\omega_n s + {\omega_n}^2\right) \tag{13}$$

where β, γ, α, ζ and ω_n represent the open loop gain, open loop pole, third pole location, desired damping ratio and desired natural frequency, respectively.

By comparing Eqs (11) and (13), the PID parameters (K_p, K_i and K_d) are

$$K_p = \frac{2\alpha{\omega_n}^2\zeta^2 + {\omega_n}^2 + \gamma^2}{\beta} \tag{14}$$

$$K_i = \frac{\alpha{\omega_n}^3\zeta}{\beta} \tag{15}$$

$$K_d = \frac{2\zeta\omega_n + \alpha\zeta\omega_n}{\beta} \tag{16}$$

To achieve a fast positioning with low overshoot performance, the design specifications are set as: settling time, t_s = 0.5 s, percentage of overshoot, $\%OS < 10\%$ and third pole location, α = 10. After calculated the PID controller parameters, the derivative gain, K_d and some portion of the proportional gain, K_p are moved to the feedback path for acquiring the PI-PD control in enhancing the transient response. Even though both PI-PD and conventional PID controllers show an approximately similar closed loop characteristic equation, both of them comprised of different control law

$$U_{PID}(s) = \left(K_p + \frac{K_i}{s} + K_d s\right)E(s) \tag{17}$$

$$U_{PI-PD}(s) = \left(K_p + \frac{K_i}{s}\right)E(s) - (K_{pb} + K_d s)X(s) \tag{18}$$

Based on Eqs (17) and (18), the conventional PID controller is functioned based on the error signal, $E(s)$ only, whereas the PI-PD controller is operated based on the error signal, $E(s)$ and the output signal $X(s)$. Hence, the PI-PD controller tends to act faster than the conventional PID controller to compensate the error.

3.2.2 Model-based feedforward control

The model-based feedforward control is employed to improve the overshoot reduction characteristic of the PI-PD control. The control law of the model-based feedforward control is expressed as

$$U_{ff}(s) = K_{ff}X_r(s) \tag{19}$$

where K_{ff} and $X_r(s)$ represent the linearized feedforward gain and reference input.

From Eq. (19), the model-based feedforward control is acted based on the desired output or reference input. Hence, by using the feedforward control, the desired output is known in advance and it can synthesize an adequate control signal to the closed loop system for moving the mechanism to the targeted output. Thus, the model-based feedforward control is used to enhance the system following characteristic and provide a better overshoot reduction characteristic. It also leads to a faster positioning time.

To design the model-based feedforward control, the relationship between the controlled voltage and the levitation displacement is obtained via experiments. First, a ramp input voltage with gradient, m = 0.1 V/t is applied to the system at different levitation displacement from 0 mm to 15 mm with every 1 mm incremental displacement. Then, the minimum voltage to levitate the steel ball at various displacements is determined. The quantitative comparisons of ten (10) repeatability

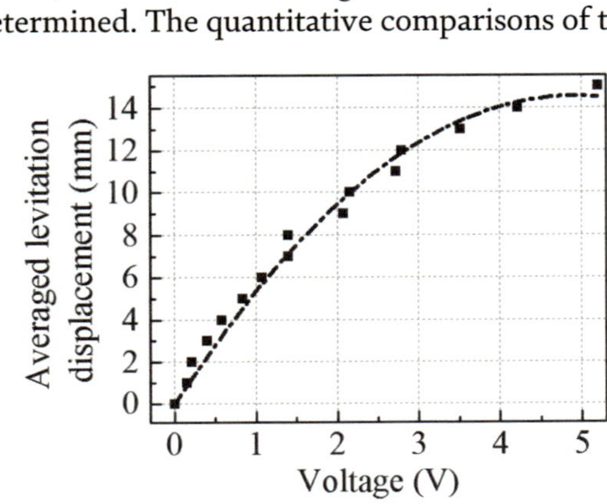

Figure 4.
The maglev system driving characteristic in open loop.

tests are carried out at various levitation displacements. **Figure 4** depicts the system driving characteristic that is measured and adopted as a model-based feedforward control in the proposed controller.

Figure 5 shows the step responses of the PI-PD and FF PI-PD controllers at 0.5 mm and 1.0 mm step inputs. In contrast to the PI-PD controller, the FF PI-PD controller demonstrates a better overshoot reduction characteristic. Besides, the FF PI-PD controller positioning time is shorter than the PI-PD controller. The comparative experimental performances show that the model-based feedforward control improves the overshoot reduction characteristic.

3.2.3 Disturbance compensator

In order to enhance the disturbance rejection characteristic of the proposed controller, a disturbance compensator is designed and incorporated with the FF PI-PD control, via lowering the magnitude of sensitivity function. In practical, the external disturbance and parameter uncertainties are lumped as an equivalent disturbance. A simple way to attenuate the equivalent disturbance is through introducing a cancelation term to it. The proposed disturbance compensator considers the difference between the actual output and the reference input as an equivalent disturbance. Then, an adequate voltage is applied to suppress the equivalent disturbance. The control law of the disturbance compensator is expressed as

$$U_z(s) = [X(s) - X_r(s)]K_{ff} \tag{20}$$

where $X(s) = K_z I(s)$, K_{ff}, K_z, $X(s)$ and $X_r(s)$ represent the linearized feedforward gain, linearized disturbance compensator gain, levitation height and reference input.

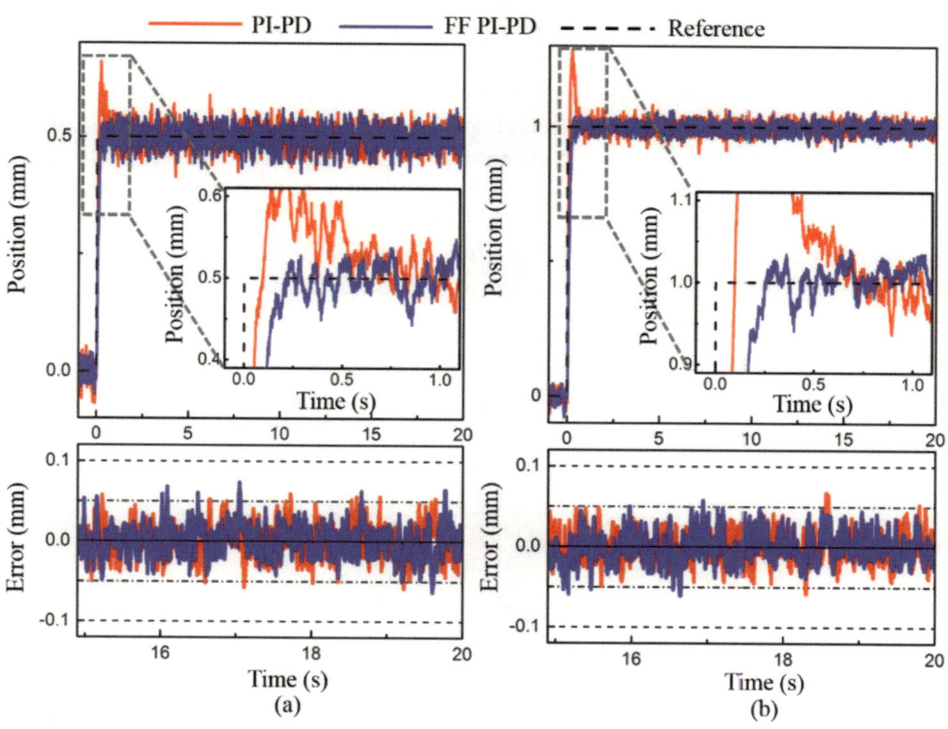

Figure 5.
Experimental step responses of the FF PI-PD and PI-PD control system. (a) Responses to a 0.5 mm step input. (b) Responses to a 1.0 mm step input.

From Eq. (20), the difference between the actual output and the reference input is considered as an estimated disturbance. Then, a sufficient voltage is applied to the control signal to attenuate the estimated disturbance.

To design the disturbance compensator, the relationship between the controlled current and the levitation displacement is attained experimentally. First, the mechanism is stabilized by a control system. Next, the required current to levitate the steel ball at different levitation displacement from −2.5 mm to 2.5 mm with every 0.5 mm interval displacement is measured. The quantitative comparisons of ten (10) repeatability tests are conducted at various levitation displacements. **Figure 6** shows the system current dynamic where 0 mm denotes the system datum or initial position which is at 10.5 mm. The system current dynamic is employed as a disturbance compensator in the proposed controller.

Figure 7 depicts the experimental impulse disturbance rejection performance of the FF PI-PD and FF PI-PD + K_z controllers. It can be seen clearly that the FF PI-PD + K_z controller is less sensitive to the external disturbance. The disturbance rejection characteristic of the FF PI-PD + K_z controller is proven theoretically by using the closed loop sensitivity function. The sensitivity functions of the FF PI-PD and FF PI-PD + K_z controllers are

$$S_{FFPI-PD}(s) = \frac{1}{1 + \left(K_p + \frac{K_i}{s} + K_{pb} + K_d \frac{s\omega_c}{s+\omega_c}\right) G_p(s)} \tag{21}$$

$$S_{FFPI-PD+K_z}(s) = \frac{1}{(K_a - K_{ff}K_z) + \left(K_p + \frac{K_i}{s} + K_{pb} + K_d \frac{s\omega_c}{s+\omega_c}\right) G_p(s)} \tag{22}$$

From Eqs (21) and (22), the proposed controller consists of the additional elements to reduce sensitivity of the system and hence to accomplish better robustness to disturbance. **Figure 8** presents the frequency responses of the FF PI-PD and FF PI-PD + K_z from the disturbance to the displacement. To decrease the effect of disturbance, the sensitivity functions of the closed-loop system must have sufficiently low magnitude. As can be seen clearly in **Figure 8**, the proposed controller consists of lower sensitivity magnitude than the FF PI-PD controller up to the range of 70 Hz. Thus, it can be expressed that the disturbance compensation control scheme of the proposed controller tends to improve the system disturbance rejection characteristic. In short, the FF PI-PD + K_z control system is less sensitive to the external disturbance and parameter variation in comparison to the FF PI-PD controller.

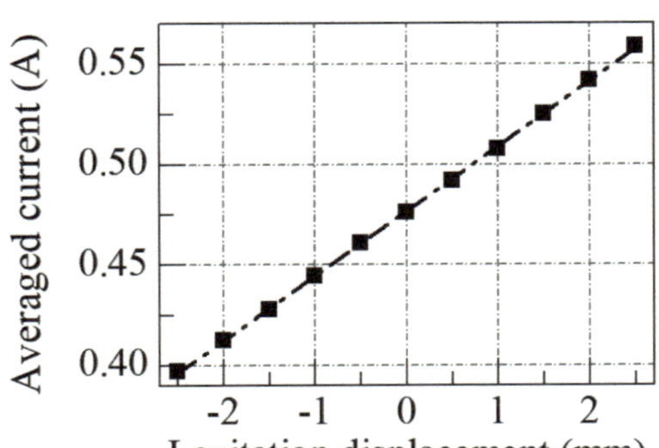

Figure 6.
The maglev system current dynamic in closed loop.

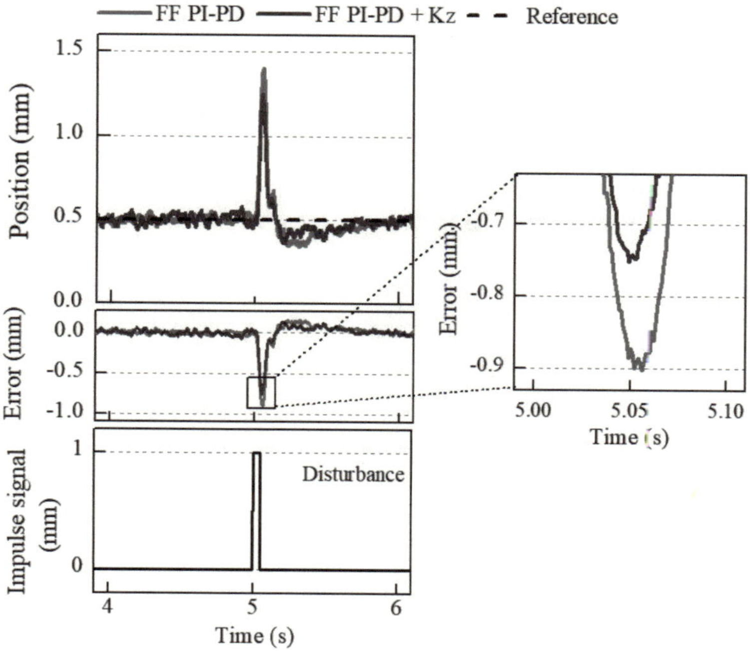

Figure 7.
Experimental impulse disturbance rejection performance.

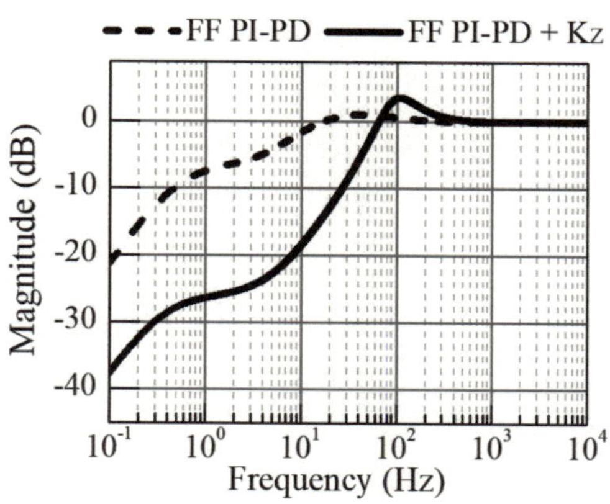

Figure 8.
Simulated frequency response for sensitivity of FF PI-PD and FF PI-PD + Kz controllers.

3.3 Stability analysis

Basically, digital control systems are used for motion control. Thus, the stability of the FF PI-PD + K_z control system is discussed in discrete-time. The nonlinear elements of the controller are undergone linearization and the linearized gains are used for the stability analysis. After linearized, the feedforward and disturbance compensator gains are 0.26 V/mm and 30.67 mm/A, respectively. The stability analysis using the linearized model is adequate to provide the important knowledge of stability. The discrete-time FF PI-PD + K_z control system is illustrated in **Figure 9**.

Using backward difference rule, the pulse transfer function of the FF PI-PD + K_z control system is expressed as

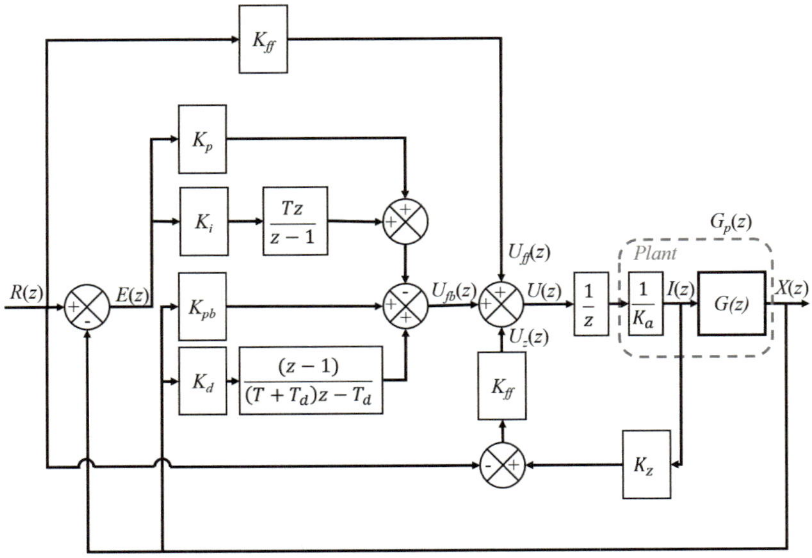

Figure 9.
Discrete-time of FF PI-PD + Kz *control system.*

$$T(z) = \frac{X(z)}{R(z)} = \frac{z^{-1}G_p(z)G_c(z)}{1 + z^{-1}G_p(z)G_c(z)} \tag{23}$$

where

$$G_p(z) = \frac{\mu_2 z + \mu_3}{z^2 - \mu_4 z + 1}$$

$$G_c(z) = \frac{\mu_5 z^2 + \mu_6 z + \mu_7}{\mu_8 z^2 - \mu_9 z + \mu_{10}}$$

$\mu_1 = \beta/\gamma^2$, $\mu_2 = \mu_1\left(e^{\gamma T} + e^{-\gamma T}\right) - \mu_1 - \frac{\mu_1}{2}\left(e^{\gamma T} + e^{-\gamma T}\right)$, $\mu_3 = -\mu_1 + \frac{\mu_1}{2}e^{\gamma T} + \frac{\mu_1}{2}e^{-\gamma T}$, $\mu_4 = e^{\gamma T} + e^{-\gamma T}$, $\mu_5 = K_i T^2 + \left(K_p - K_{pb} + K_{ff}K_z\right)T + \left(K_p - K_{pb} + K_{ff}K_z\right)T_d + K_i T T_d - K_d$, $\mu_6 = \left(K_{pb} - K_{ff}K_z - K_p\right)T + \left(2K_{pb} - 2K_{ff}K_z - 2K_p\right)T_d - K_i T T_d + 2K_d$, $\mu_7 = \left(K_p + K_{ff}K_z - K_{pb}\right)T_d - K_d$, $\mu_8 = (T + T_d)$, $\mu_9 = T + 2T_d$, $\mu_{10} = T_d$ and $T_d = 1.60 \times 10^{-3}$ s.

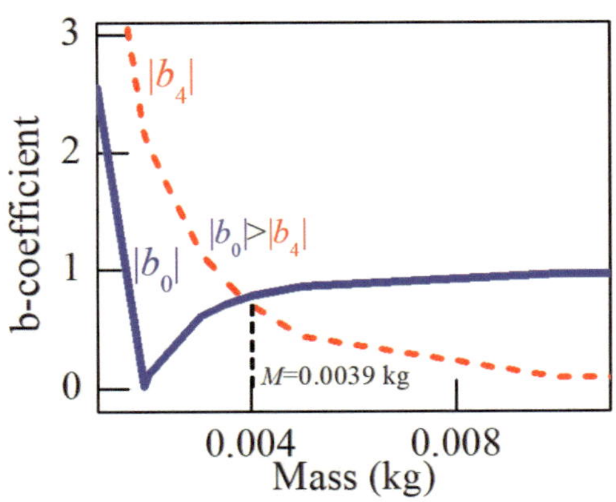

Figure 10.
b-*coefficient of mass parameter variation.*

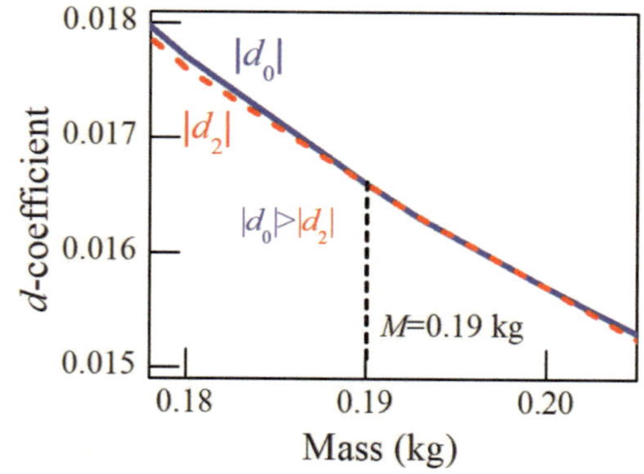

Figure 11.
d-*coefficient of mass parameter variation.*

The Jury stability test is used to examine the stability limit of the FF PI-PD + K_z control system. The characteristic in Eq. (23) is used to identify the stability limit of the FF PI-PD + K_z control system. **Figures 10** and **11** show the minimum and maximum values of the mass parameter variation, respectively. The results show that the parameter that influences the stability of the control system is the object mass, *M*. To maintain the system stable, the object mass, *M* must be kept between 3.9 g < *M* < 190 g. In short, the Jury test proves that the FF PI-PD + K_z control system remains stable with the increment of mass weight to two times of its default one.

4. Experimental performance

Experiments are conducted to evaluate the effectiveness of the FF PI-PD + K_z control system. Two types of motion control that are positioning and tracking controls are experimentally examined. The full-state feedback (FSF) controller is designed and compared with the proposed control. FSF control is chosen for the comparison purpose is because it is an advanced controller that has been regularly applied to the nonlinear applications such as maglev system [18, 19], inverted pendulum system [20] and others. Besides, the FF PI-PD control is compared with the proposed one in order to prove the usefulness of the disturbance compensator. The robust performance of the proposed control is examined by injecting an impulse disturbance to the system, and followed by increasing the mass of the ball by 25%.

Figure 12 illustrates the block diagram of the FSF controller. An integral action is added into the FSF controller to eliminate the steady state error by increasing the transfer function to type one system. From **Figure 12**, the system state-space model is written as

$$\begin{bmatrix} \dot{x}(t) \\ \dot{e}(t) \end{bmatrix} = \begin{bmatrix} A & 0 \\ -C & 0 \end{bmatrix} \begin{bmatrix} x(t) \\ e(t) \end{bmatrix} + \begin{bmatrix} B \\ 0 \end{bmatrix} u(t) + \begin{bmatrix} 0 \\ 1 \end{bmatrix} r(t) \tag{24}$$

where.

$$\mathrm{A} = \begin{bmatrix} 0 & 1 & 0 \\ K_x/M & 0 & -K_c/M \\ 0 & 0 & 0 \end{bmatrix}, \quad \mathrm{B} = \begin{bmatrix} 0 \\ 0 \\ 1/K_a \end{bmatrix} \text{ and } \mathrm{C} = [K_s \;\; 0 \;\; 0]$$

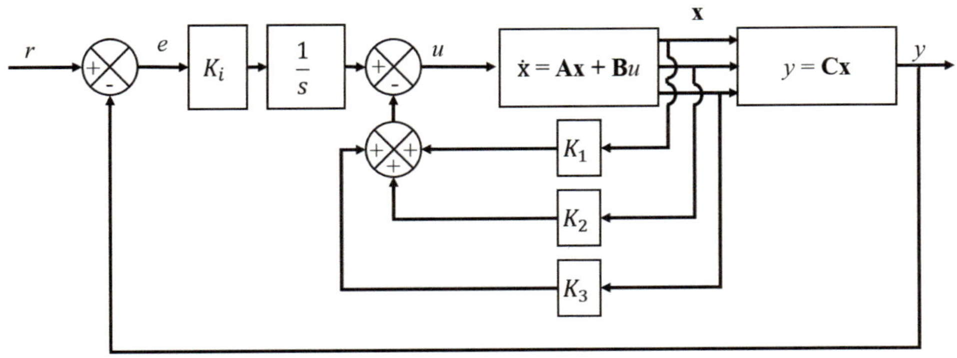

Figure 12.
Block diagram of the full-state feedback (FSF) controller.

Based on Eq. (24), the state feedback control law, $u_{fsf}(t)$ is defined as

$$u_{fsf}(t) = -\mathbf{K}\mathbf{x}(t) + K_i e(t) \tag{25}$$

where $\mathbf{K}$, K_i, $x(t)$ and $e(t)$ represent the state feedback gain matrix, integral gain, levitation displacement and error, accordingly.

Ackermann's formula is used to determine the state feedback gain matrix, $\mathbf{K}$ and the integral gain, K_i. For comparative purpose, the design specifications of FSF controller are set as: settling time, t_s = 0.5 s, percentage of overshoot, $\%OS < 10\%$ as well as third and fourth poles location, α = 10. The frequency of first-order low-pass filter, ω_c is selected based on the system cut-off frequency at around 600 rad/s. The FSF controller parameters are tuned to have the best positioning performance at 1.0 mm step response as similar to the FF PI-PD + K_z controller (see **Figure 13**). **Table 2** shows the controller parameters for FSF, FF PI-PD and FF PI-PD + K_z control systems.

4.1 Positioning performance

In this experiment, the initial position is set at 10.5 mm and the working range is within ±2.5 mm. **Figures 14** and **15** show the experimental positioning performance of the FSF, FF PI-PD and FF PI-PD + K_z control systems to 0.5 mm, 1.0 mm, −0.5 mm and − 1.0 mm step inputs, respectively. As observed clearly, the FF PI-PD + K_z controller shows almost identical positioning performance, with no overshoot as the FF PI-PD and FSF control systems. However, the FSF control system

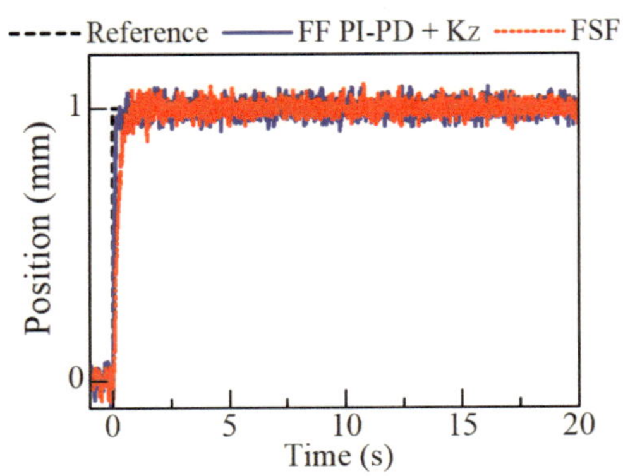

Figure 13.
Experimental step responses of controllers at 1.0 mm reference input.

Controller	K_1	K_2	K_3	K_p	K_i	K_{pb}	K_d
FSF	0.80	0.02	0.01	—	2.36	—	—
FF PI-PD	—	—	—	0.45	1.20	0.15	0.03
FF PI-PD + K_z	—	—	—	0.45	1.20	0.15	0.03

Table 2.
Controller parameters.

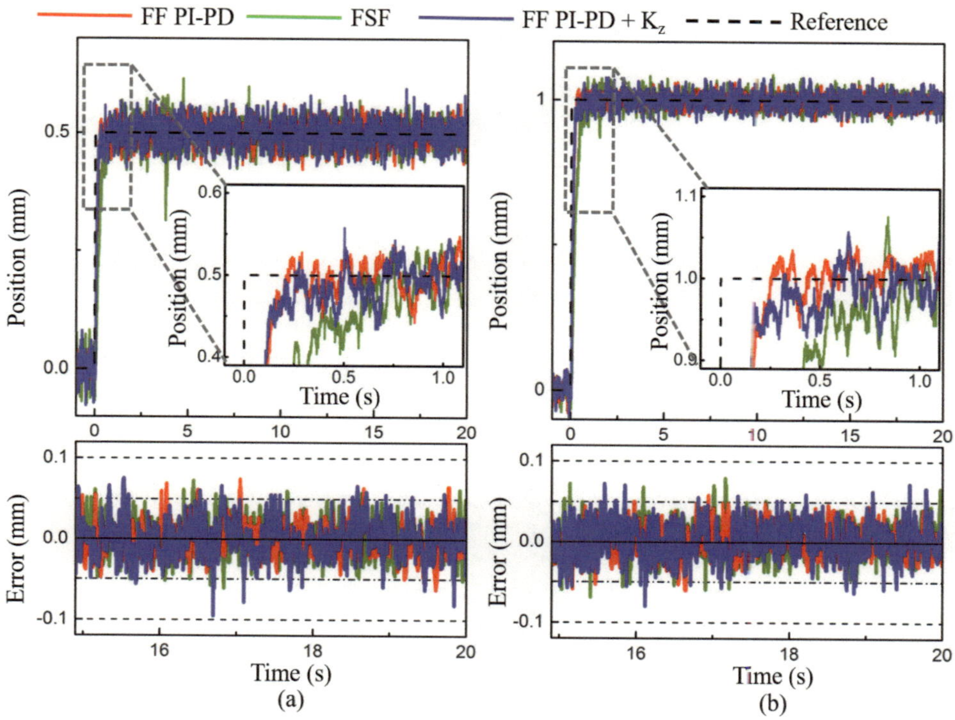

Figure 14.
Experimental step responses of the three control systems at positive side direction. (a) Responses to a 0.5 mm step input (default mass). (b) Responses to a 1.0 mm step input (default mass).

takes longer settling time than the FF PI-PD and FF PI-PD + K_z controllers to reach steady-state that less than ±100 μ m. **Figure 16** presents the simulated closed-loop frequency response for the three control systems. As can be seen in **Figure 16**, the FF PI-PD and FF PI-PD + K_z controls demonstrate wider bandwidth as compared to the FSF control. Therefore, it can be explained that the both FF PI-PD and FF PI-PD + K_z controllers could perform shorter settling time than the FSF controller.

Table 3 shows the quantitative comparison of twenty (20) repeatability experimental results for the point-to-point motion of the three controllers. The settling time, t_s is determined as the time where the system is stabilized within ±100 μm. All three controllers demonstrate zero overshoot at every step input. Although FSF performs zero overshoot in all the step input, it takes long settling time to reach steady state that of less than ±100 μm. The settling time of the FF PI-PD + K_z is 65.6% shorter than the FSF controller.

4.2 Tracking performance

For tracking motion, periodic trapezoidal reference input is utilized to command the maglev system. The maximal tracking error is stated as E_{max} = max $|x_r - x|$ where

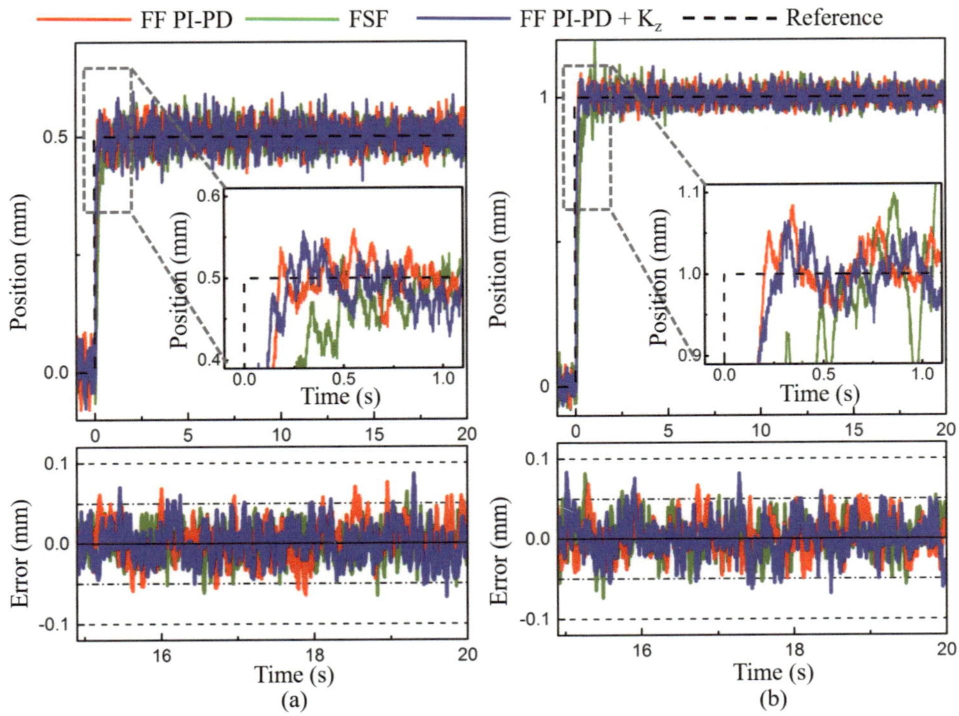

Figure 15.
Experimental step responses of the three control systems at negative side direction. (a) Responses to a − 0.5 mm step input (default mass). (b) Responses to a −1.0 mm step input (default mass).

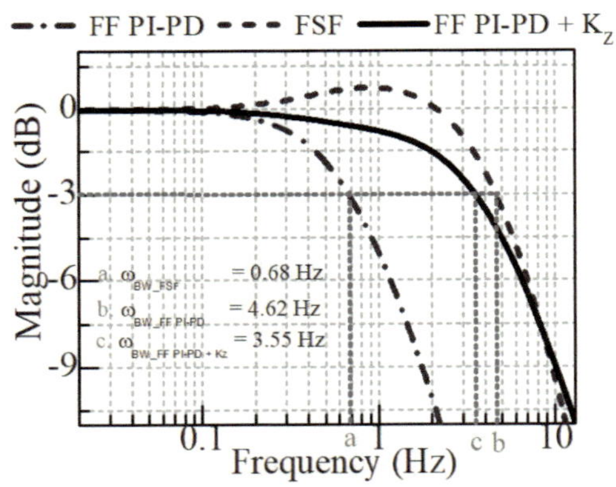

Figure 16.
Simulated closed-loop frequency response.

x_r is the reference input and x is the levitation height. In addition, the root mean square error, E_{rms} is calculated as $\sqrt{1/N\sum_{k=1}^{N}e^2}$ where N represents the number of data samples and e is the tracking error.

Figure 17 illustrates the trapezoidal tracking performance of the FSF, FF PI-PD and FF PI-PD + K_z control systems to 0.5 mm and 1.0 mm amplitudes. Both FF PI-PD and FF PI-PD + K_z controllers demonstrate almost identical tracking performance. The tracking error difference between them is insignificant. On the other hand, the FSF controller demonstrates the worst tracking performance with the largest tracking error among the compared controllers. The maximum tracking

Step height	Performance index		FF PI-PD + K_z	FF PI-PD	FSF
0.5 mm	*OS*, %	Average	0.00	0.00	0.00
		Standard deviation	0.00	0.00	0.00
	t_s, s	Average	1.45×10^{-1}	1.17×10^{-1}	4.35×10^{-1}
		Standard deviation	3.01×10^{-2}	1.91×10^{-2}	6.37×10^{-2}
1.0 mm	*OS*, %	Average	0.00	0.00	0.00
		standard deviation	0.00	0.00	0.00
	t_s, s	Average	2.02×10^{-1}	1.60×10^{-1}	5.48×10^{-1}
		Standard deviation	3.82×10^{-2}	1.19×10^{-2}	5.45×10^{-2}
−0.5 mm	*OS*, %	Average	0.00	0.00	0.00
		Standard deviation	0.00	0.00	0.00
	t_s, s	Average	1.25×10^{-1}	1.90×10^{-1}	3.83×10^{-1}
		Standard deviation	3.22×10^{-2}	1.19×10^{-1}	8.64×10^{-2}
−1.0 mm	*OS*, %	Average	0.00	0.00	0.00
		Standard deviation	0.00	0.00	0.00
	t_s, s	Average	1.77×10^{-1}	2.03×10^{-1}	5.06×10^{-1}
		Standard deviation	2.35×10^{-2}	6.07×10^{-2}	5.36×10^{-2}

OS: *overshoot*, t_s: *settling time.*

Table 3.
Experimental positioning performances of twenty (20) experiments for three controllers (default mass).

error of the FSF controller at 0.5 mm amplitude is around 1.5 times larger than the FF PI-PD + K_z controller. Meanwhile, as the amplitude increased to 1.0 mm, the FSF controller maximum tracking error is about 2 times larger than the FF PI-PD + K_z controller (see error signal in **Figure 17(b)**). The maximum tracking error occurred at the slope of the trapezoidal signal where the velocity changes. Thus, the experimental results proved that the FSF controller has low adaptability to the velocity change. The FSF controller comprised of narrow bandwidth (see **Figure 16**). Hence, it can explain that the FSF controller does not have sufficient speed to cope with the variation of velocity effectively. The average of E_{max} and E_{rms} values of twenty (20) experiments for the tracking motion is summarized in **Table 4**. At amplitude 0.5 mm, the E_{max} and E_{rms} values of the FSF controller are 48.2% and 46.7% larger than the FF PI-PD + K_z controller. Besides, the E_{max} and E_{rms} values of the FF PI-PD + K_z controller are 47.1% and 58.9% smaller than the FSF controller at 1.0 mm amplitude. On the other hand, the difference of E_{max} and E_{rms} values between the FF PI-PD and the FF PI-PD + K_z controllers are insignificant. Overview, the FF PI-PD and FF PI-PD + K_z control systems track the trapezoidal signal more accurately and precisely with the smaller E_{max} and E_{rms} values as compared to the FSF controller.

4.3 Robustness performance

The robust performance of the proposed controller is evaluated in the presence of mass variation. The 25% extra load is added to the default load of the mechanism. In this experiment, the control performance is examined in two type of motions: point-to-point and tracking motions. The robust performance of the FF PI-PD + K_z controller is then compared with the FF PI-PD and FSF controllers.

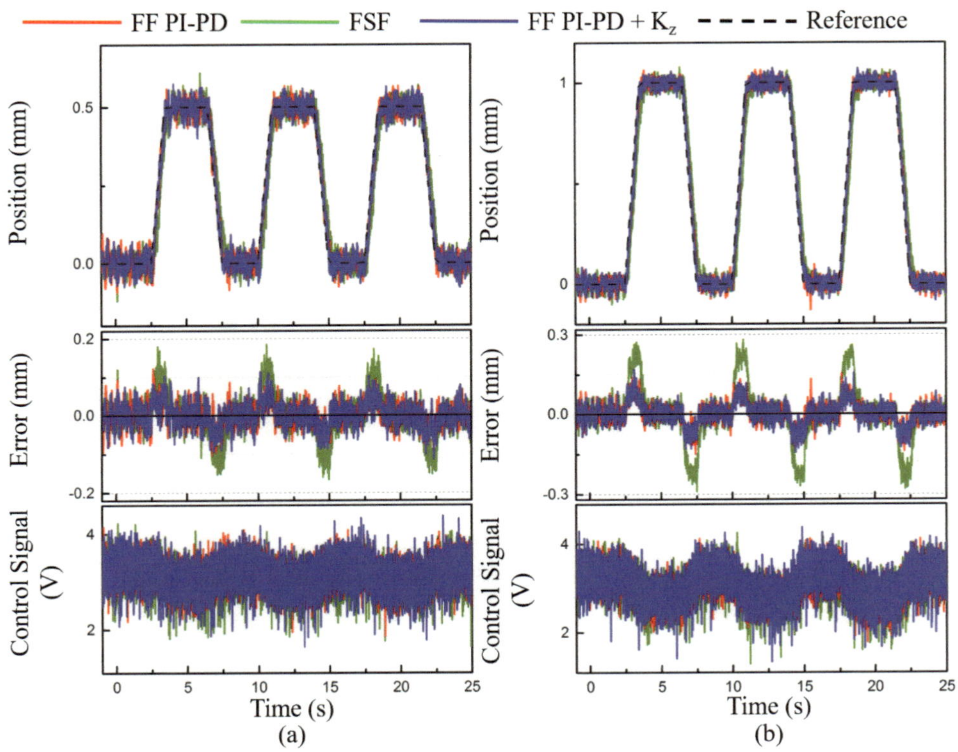

Figure 17.
Comparative experimental trapezoidal tracking responses of the three controllers. (a) Responses to a trapezoidal input: 0.5 mm (default mass). (b) Responses to a trapezoidal input: 1.0 mm (default mass).

Reference input	Controller	E_{max}	E_{rms}
		Average, mm	**Average, mm**
Trapezoidal, 0.5 mm	FSF	2.28×10^{-1}	5.72×10^{-2}
	FF PI-PD	1.21×10^{-1}	3.04×10^{-2}
	FF PI-PD + K_z	1.18×10^{-1}	3.05×10^{-2}
Trapezoidal, 1.0 mm	FSF	2.95×10^{-1}	1.04×10^{-1}
	FF PI-PD	1.51×10^{-1}	4.24×10^{-2}
	FF PI-PD + K_z	1.56×10^{-1}	4.27×10^{-2}

Table 4.
Average of twenty (20) experiments trapezoidal motion for the three controllers (default mass).

4.3.1 Point-to-point motion

The positioning responses with the increased mass are shown in **Figures 18** and **19**. As the mass is increased, the FF PI-PD controller shows overshoot occurrence at both positive and negative directions. Thus, the FF PI-PD controller fails to demonstrate its robust performance. On the other hand, the FF PI-PD + K_z controller demonstrates high robustness via demonstrating zero overshoot at all the step responses regardless of the variation of mass. Hence, the experimental positioning results proved that the disturbance compensation control scheme is comprised in the FF PI-PD + K_z controller and it has led to the less sensitive to parameter variation characteristic of the controller. Although the FSF controller performs its good robustness through showing

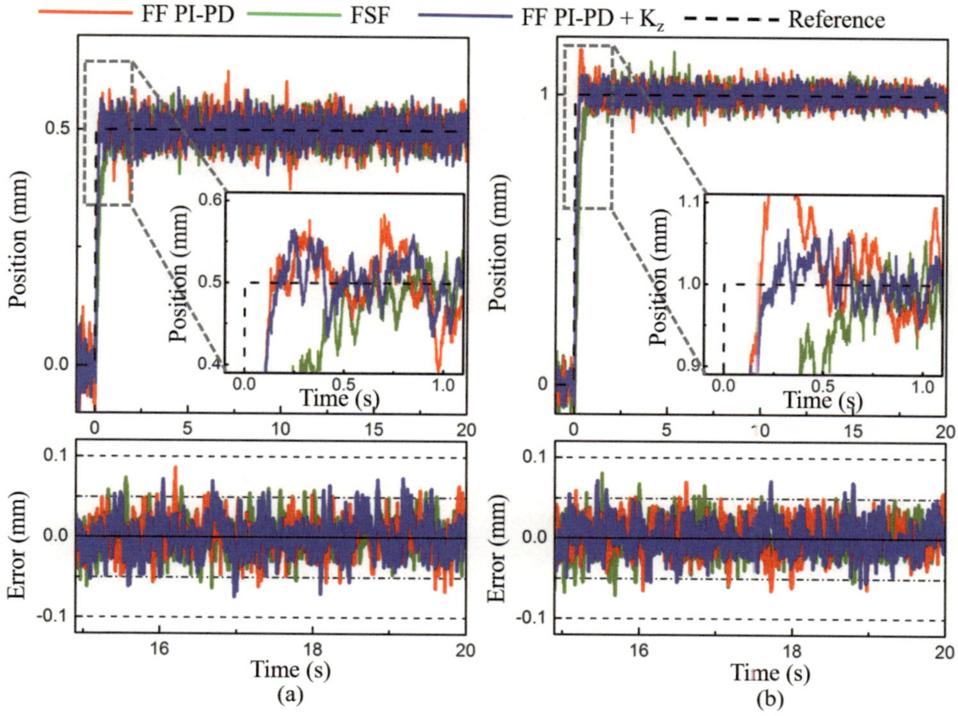

Figure 18.
Experimental step responses of the three control systems at positive side direction. (a) Responses to a 0.5 mm step input (increased mass). (b) Responses to a 1.0 mm step input (increased mass).

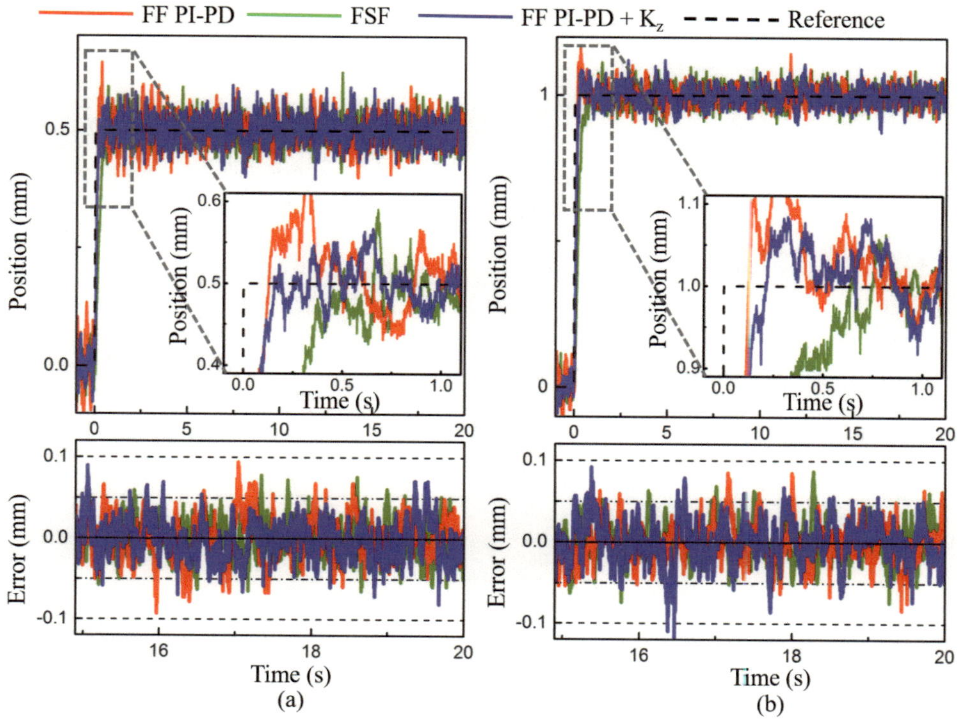

Figure 19.
Experimental step responses of the three control systems at negative side direction. (a) Responses to a −0.5 mm step input (increased mass). (b) Responses to a −1.0 mm step input (increased mass).

zero overshoot at all the step responses, it takes longer positioning time than the FF PI-PD + K_z controller to reach the steady-state (**Table 5**).

Table 6 shows the quantitative comparison of twenty (20) repeatability tests for the point-to-point motion in the presence of mass variation. As can be seen from **Table 6**, when the mass of table is increased, the FF PI-PD controller fails to demonstrate its robustness by producing a large overshoot. The change of mass has caused the overshoot of the FF PI-PD controller is increased by 20% of the default mass condition and the settling time of the FF PI-PD controller is 47.9% longer than the FF PI-PD + K_z controller. In contrast, the FF PI-PD + K_z controller has successfully remained its high robust performance via demonstrating zero overshoot at all the step responses. It is evident that the FF PI-PD + K_z controller enhances the robustness of the FF PI-PD controller via introducing the disturbance compensation

Step height	Performance index		FF PI-PD + K_z	FF PI-PD	FSF
0.5 mm	*OS*, %	Average	0.00	0.00	0.00
		Standard deviation	0.00	0.00	0.00
	t_s, s	Average	1.13×10^{-1}	1.77×10^{-1}	3.92×10^{-1}
		Standard deviation	1.60×10^{-2}	1.05×10^{-1}	1.28×10^{-1}
1.0 mm	*OS*, %	Average	0.00	1.40×10^{1}	0.00
		Standard deviation	0.00	2.52×10^{-2}	0.00
	t_s, s	Average	1.50×10^{-1}	2.95×10^{-1}	5.08×10^{-1}
		Standard deviation	1.47×10^{-2}	9.75×10^{-2}	8.60×10^{-2}
−0.5 mm	*OS*, %	Average	0.00	2.80×10^{1}	0.00
		Standard deviation	0.00	2.85×10^{-2}	0.00
	t_s, s	Average	1.15×10^{-1}	2.77×10^{-1}	3.59×10^{-1}
		Standard deviation	5.40×10^{-2}	1.93×10^{-1}	8.42×10^{-2}
−1.0 mm	*OS*, %	Average	0.00	1.75×10^{1}	0.00
		Standard deviation	0.00	3.66×10^{-2}	0.00
	t_s, s	Average	2.27×10^{-1}	4.32×10^{-1}	4.57×10^{-1}
		Standard deviation	1.17×10^{-2}	4.67×10^{-2}	6.87×10^{-2}

OS: *overshoot,* t_s*: settling time.*

Table 5.
Experimental positioning performances of twenty (20) experiments for three controllers (increased mass).

Reference input	Controller	E_{max}	E_{rms}
		Average, mm	**Average, mm**
Trapezoidal, 0.5 mm	FSF	1.92×10^{-1}	5.33×10^{-2}
	FF PI-PD	1.53×10^{-1}	3.50×10^{-2}
	FF PI-PD + K_z	1.26×10^{-1}	3.05×10^{-2}
Trapezoidal, 1.0 mm	FSF	2.99×10^{-1}	9.87×10^{-2}
	FF PI-PD	1.72×10^{-1}	4.04×10^{-2}
	FF PI-PD + K_z	1.57×10^{-1}	3.84×10^{-2}

Table 6.
Average of twenty (20) experiments trapezoidal tracking motion for three controllers (increased mass).

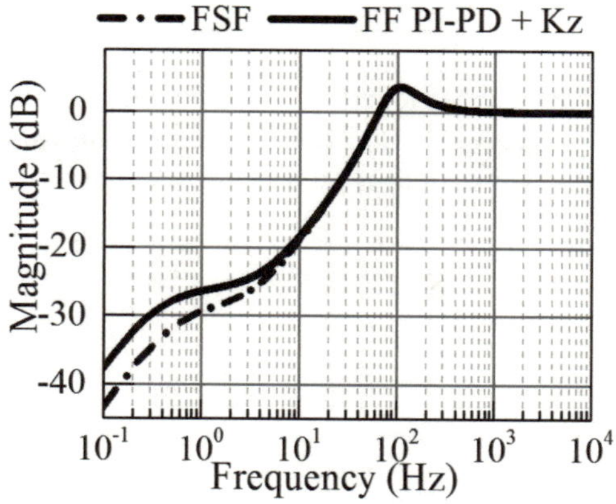

Figure 20.
Sensitivity response of the FF PI-PD + K_z and the FSP controllers.

control scheme. On the other hand, the FF PI-PD + K_z achieves approximately three (3) times shorter settling time than the FSF controller when the mass increased. In short, the FF PI-PD + K_z controller demonstrates the best positioning performance among the compared controllers regardless of the mass variation.

As referred to the **Figure 20**, the FF PI-PD + K_z controller performs smaller sensitivity function magnitude than the FF PI-PD controller. Hence, it is less sensitive to parameter variation. In short, despite the variation of mass and amplitude, the FF PI-PD + K_z controller demonstrates a superior tracking performance than the FF PI-PD and FSF controllers, where it tracks the trapezoidal command accurately and precisely via illustrating the lowest E_{max} and E_{rms} values.

5. Conclusions

In this chapter, the architecture of the FF PI-PD + K_z control system for enhancing the positioning, tracking and robust performances of the maglev system is presented. Initially, a two-degree-of-freedom (2 DOF) PID control – PI-PD, is used to improve the transient response of the conventional PID controller by minimizing the resonance peak. However, the PI-PD control has not sufficiently performed promising positioning responses. A as solution, a model-based feedforward (FF) control is integrated to the PI-PD control for further improving the following characteristic and overshoot reduction capabilities of the mechanism. Lastly, a disturbance compensator (K_z) is served to enhance the system robustness via lowering the sensitivity function magnitude. Although the framework of proposed controller - FF PI-PD + K_z control system is slightly complex than the conventional PID controller, but the design procedure of FF PI-PD + K_z control system remains simple, straightforward, and ease to understand. This advantageous highlight the applicability of the FF PI-PD + K_z control system in the industrial applications. The effectiveness of the proposed controller is evaluated experimentally in point-to-point and tracking motions in comparison to the FF PI-PD and Full State Feedback (FSF) controllers. The robust performance of the controllers is examined in the presence of the mass variations. As an overview, the FF PI-PD + K_z control system performs well in the positioning and robustness performances as compared to the FF PI-PD and FSF controllers. The comparative experimental results are sufficient to prove the contribution of the FF PI-PD + K_z control system in overshoot reduction and robustness enhancement. As for future work, the robustness performance, and the positioning accuracy of the FF PI-PD + K_z control system will be improved.

Acknowledgements

The authors would like to be obliged to Motion Control Research Laboratory, Center for Robotics and Automations, Faculty of Electrical Engineering, Universiti Teknikal Malaysia Melaka for providing the laboratory facilities and equipment support.

Nomenclature

M	steel ball mass
N_w	number of windings
x_o	nominal displacement
i_o	nominal current
K	electromagnetic constant
K_a	power amplifier
K_s	sensor sensitivity
g	gravitational acceleration
F_m	electromagnetic force
F_g	gravitational force
i	current
x	displacement
g	gravitational acceleration
K_c	current coefficient
K_x	position coefficient
β	open loop gain
γ	open loop pole
α	third pole location
ζ	desired damping ratio
ω_n	desired natural frequency
K_p	proportional gain
K_i	integral gain
K_d	derivative gain
K_{pb}	feedback proportional gain
K_{ff}	linearized feedforward gain
K_z	linearized disturbance compensation gain
δ_{PID}	closed loop PID control characteristic equation
$\delta_{PI\text{-}PD}$	closed loop PI-PD control characteristic equation
$\delta_{desired}$	desired characteristic equation
G_p	plant model
G_{ff}	model-based feedforward control
$G_{PI\text{-}PD}$	PI-PD control
G_z	disturbance compensator
T	sampling time
T_d	time constant of derivative elements with filter
ω_c	system cutoff frequency
$\mathbf{K}$	state feedback gain matrix
u_{fb}	feedback control signal
u_{ff}	feedforward control signal
u	summation signals of u_{fb} and u_{ff}
N	number of data sample
e	tracking error

x_r	reference input
E_{max}	maximal error
E_{rms}	root mean square error

Author details

Shin-Horng Chong*, Roong-Soon Allan Chan and Norhaslinda Hasim
Faculty of Electrical Engineering, Centre for Robotics and Industrial Automation, Universiti Teknikal Malaysia Melaka, Durian Tunggal, Malaysia

*Address all correspondence to: horng@utem.edu.my

References

[1] Yaghoubi H. The most important maglev applications. Journal of Engineering. 2013; 1:19. DOI: https://doi.org/10.1155/2013/537986

[2] Banerjee S, Prasad D, Pal J. Large gap control in electromagnetic levitation. The Instrumentation, Systems, and Automation Society. 2006;45(2): 215-224. DOI: https://doi.org/10.1016/S0019-0578(07)60191-8

[3] Banerjee S, Prasad D, Pal J. Design, implementation and testing of a single axis levitatin system for suspension of a platform. ISA Transactions. 2007;46(2): 239-246. DOI: https://doi.org/10.1016/j.isatra.2006.09.001

[4] Fujimoto K, Sugie T. Freedom in coordinate transformation for exactineariation and its application to transient behavior impovement. Automatica. 2001;37(1): 137-144. DOI: https://doi.org/10.1016/S0005-1098(00)00134-5

[5] Cho D, Kato Y, Darin S. Sliding mode and classical control magnetic levitation system. IEEE Control Systems. 1993;13(1): 42–48. DOI: 10.1109/37.184792

[6] Sinha PK, Pechev AN. Nonlinear H-infinity controllers for electromagnetic suspension systems. IEEE Transactions on Automatic Control. 2004;49(4): 563–568. DOI: 10.1109/TAC.2003.822865

[7] Yang J, Zalotas A, Chen WH, Michail K, Li S. Robust control of nonlinear maglev suspension system with mismatched uncertainties via DOBC approach. ISA Transactions. 2011;50(3): 389-396. DOI: https://doi.org/10.1016/j.isatra.2011.01.006

[8] Yang ZJ, Tateshi M. Adaptive robust nonlinear control of a magnetic levitation system. Automatica. 2001;37(7): 1125-1131. DOI: https://doi.org/10.1016/S0005-1098(01)00063-2

[9] Bächle T, Hentzelt S, Graichen K. Nonlinear model predictive control of a magnetic levitation system. Control Engineering Practice. 2013;21(9): 1250-1258. DOI: https://doi.org/10.1016/j.conengprac.2013.04.009

[10] Qin Y, Peng H, Zhou F, Zeng X, Wu J. Nonlinear modeling and control approach to magnetic levitation ball system using functional weight RBF network-based state dependent ARX model. Journal of the Franklin Insitute. 2015;352(10): 4309-4338. DOI: https://doi.org/10.1016/j.jfranklin.2015.06.014

[11] Klau M, Kalúz M, Kvasnica M. Real-time implementation of an explicit MPC-based reference governor for control of a magnetic levitation system. Control Engineering Practice. 2017;60: 99-105. DOI: https://doi.org/10.1016/j.conengprac.2017.01.001

[12] Golob M, Tovornik B. Modeling and control of the magnetic suspension system. ISA Transactions. 2003;42: 89-100. DOI: https://doi.org/10.1016/S0019-0578(07)60116-5

[13] Chen SY, Lin FJ, Shyu KK. Direct decentralized neural control for nonlinear MIMO magnetic levitation system. Neurocomputing. 2009;72(13): 3220-3230. DOI: https://doi.org/10.1016/j.neucom.2009.02.009

[14] Leva A, Bascetta L. Designing the feedforward part of 2-dof industrial controllers for optimal tracking. Control Engineering Practice. 2007;15(8): 909-921. DOI: https://doi.org/10.1016/j.conengprac.2006.11.009

[15] Ghosh A, Krishnan TR, Tejaswy P, Mandal A, Pradhan JK, Ranasingh S. Design and implementation of a 2-dof PID compensation for magnetic levitation systems. ISA Transactions. 2014;53(4): 1216-1222. DOI: https://doi.org/10.1016/j.isatra.2014.05.015

[16] Soon ACR, Chong SH, Said MA. 2-dof lead-plus-PI control approach for magnetic levitation system. Journal of Telecommunications, Electronic and Computer Engineering. 2016;8(11): 3–8.

[17] Soon ACR, Chong SH, Said MA. Position control of a magnetic levitation system via a PI-PD control with feedforward compensation. In: Proceedings of the 56th Annual Conference of the Society of Instrumentation and Control Engineers of Japan (SICE'17): 19–22 September 2017; Kanazawa, Japan: IEEE; 2017. p. 73–78

[18] Schuhmann T, Hofmann W, Werner R. Improving operational performance of active magnetic bearings using kalman filter and state-feedback control. IEEE Transactions on Industrial Electronics. 2012;59(2): 821–829. DOI: 10.1109/TIE.2011.2161056

[19] Wiboonjaroen W, Sujitjorn S. Stabilization of a magnetic levitation control system via state-PI feedback. International Journal of Mathematical Models and Methods in Applied Sciences. 2013;7(7): 717–727.

[20] Tang TF, Chong SH, Pang KK. Stabilization of a rotary inverted pendulum system with double PID and LQR control: experimental verification. International Journal of Automation and Control. 2020;14(1): 18-33. DOI: https://doi.org/10.1504/IJAAC.2020.103799

Chapter 6

Multi-Parameter Estimation of Uncertain Systems Based on the Extended PID Control Method

Jinping Feng and Wei Wang

Abstract

Parameter estimation is an important step in the identification of systems. With the extension of systems, there needs the multi-parameter estimation of systems. The estimation of multi parameters of complex systems based on the extended PID controllers is considered in this chapter. As the related references proved that the integral item of the nonlinear PID controller could deal with the uncertain part of the complex system (which can also be called new stripping principle, simple notes as NSP). Based on this theory, new multi-parameter estimation method is given. Firstly, the unknown parameters are expanded to new states of the system. Two cases, parameters are constant or changing with time, are separately analyzed. In the time-variant case, the unknown parameters are extended to functions which actual forms are uncertain. Secondly the method NSP could be applied to cope with the uncertain part, and then reconstruction state observation to estimate the states. If the states are observed, the unknown parameters are obtained at the same time. Finally the convergence analysis of the error systems and some simulations will be given in this chapter to indicate the effectiveness of the proposed method.

Keywords: multi-parameter, parameter estimation, complex system, extended PID control, convergence analysis

1. Introduction

Dynamic process model is the basis to study the uncertain systems. Generally speaking, the establishment of dynamic process model for the research object is the first step to solve the problem, and the parameter estimation of the established dynamic process model is the next key procedure to solve the problem. So the identification of dynamic processes is of great significance.

The design of the state observer in the control theory is to construct a dynamic system artificially, to make it approximate the real state of the dynamic system by selecting a certain form of the observer. The criterion for designing the state observer is to make the error system asymptotically converge to the origin, that is to say, as time goes by, the error will asymptotically converge to zero. It is based on this design idea we use in the parameter estimation problem. In the identification of the model, the dynamic process model is often accompanied with unknown disturbances. In the analysis of the estimation of multiple time-varying parameters, when the parameters are expanded to the states, there are also unknown parts in the dynamic process model. So this chapter will study a method for estimating multiple time-varying parameters based on the combination of disturbance stripping principle with state observer.

The reference [1] proposed a general form of establishing the state observer of the nonlinear system, and gave a direct method to deal with the nonlinear control system [2]. On the basis of the references [1–3], several specific state observers was provided to realize the estimation of a single time invariant parameter, and appropriate design parameters were selected according to the relevant results in the book [4]. By analyzing the stability the error system, a design method that made the error system asymptotically converge to zero was obtained. The simulation results showed that this method can estimate the parameters effectively [3].

For the estimation of time-varying parameters, the article [5] analyzed a system with one time-varying parameter. The design of the state observer in this article used the combination of binary control with PID control, which can handle the unknown items in the extended states. Although there was no rigorous theoretical proof in this article, the effect of parameter estimation did have excellent characteristics of fast convergence with less chatter. The reference [6] gave a method of combining binary control with nonlinear PID controller, and conducted a rigorous theoretical proof. Then it was extended to the regulation of high-level systems, and the principle of disturbance stripping [7] for the regulation of complex network systems. This laid the foundation for the theoretical analysis of the estimation methods of multiple time-varying parameters below. So this chapter is based on [5–7] and other references. The method of estimating a time-varying parameter in the nonlinear system in [5] is extended to the estimation of multiple time-varying parameters in a dynamic system by using the principle of disturbance stripping in the article [7]. The simulation studies showed that this method was also suitable for the estimation of time-invariant parameters.

The content of this chapter is arranged as follows: The Section 2 simply introduces the main idea of NSP and gives detail proof of it. The Section 3 puts forward an estimation method that contains multiple time-varying parameters in a nonlinear system. It describes the applicable objects of this kind of parameter estimation method, and gives a design of a specific state observer. Theoretical analysis and simulation research verifies the feasibility of the method. Section 4 summarizes the research methods and results presented in this chapter.

2. The main idea of NSP

The PID control method applies the error $\epsilon(t)$ between the reference input and observation. The PID control is the linear combination of the error, its differential and its integration. That is

$$u(t) = k_P\epsilon(t) + k_I\int_{t_0}^{t}\epsilon(\tau)d\tau + k_D\dot{\epsilon}(t) \tag{1}$$

where k_P, k_I, k_D are design parameters, $\epsilon(t)$ is the error, $\dot{\epsilon}(t)$ is the differential of the error, $\int_{t_0}^{t}\epsilon(\tau)d\tau$ is the integration of the error, t_0 is the intial time.

The theory analysis and large applications showed that the PID control u often had the conflict in fast and overshoot. Luckily, the nonlinear PID could solve this problem [8], which used the nonliner function such as sat funtion, fal function. It was the nonlinear combination of the error, its difference and its integration. At the same time, it also applied the nonlinear tracking-differentiator to filter the noise of the observation, and got the differential of the signal which may be not differentiable. The detail of this nonlinear PID controller can be seen in [8].

Basd on the idea of the extended PID controller, the NSP thought was proposed [6, 10, 12, 13]. They found that the integration of the error in the extended PID controller could stripping the unknown item in the complex systems. So we could use the NSP to deal with the system with unknown parts. The basic conclusion to be used in the following analysis, which is the most important thought in NSP involved in [6, 10, 12, 13]. The core idea will be simplified here, given in the form of a lemma, and with detailed proof.

Lemma 1 If the dynamic process $\mu(t)$ takes the following form:

$$\dot{\mu}(t) = \begin{cases} -\gamma sign(\sigma(t)), & |\mu(t)| \le 1, |\mu(t_0)| \le 1 \\ -\omega\mu(t), & |\mu(t)| > 1 \end{cases} \tag{2}$$

where $\sigma(t) = g(t) + k\mu(t)\int_{t_0}^{t}|e(\tau)|d\tau$, $e(t)$ is the difference between the state observer system and the original system (which is $\overline{x}$ as mentioned below), $g(t)$ is the unknown quantity with the known variation range, $\gamma > 0$, $\omega > 0$ is the undetermined constant. When the condition

$$k > \sup_{t \ge t_0}\left|\frac{g(t)}{\int_{t_0}^{t}|e(\tau)|d\tau}\right| \tag{3}$$

is satisfied, there will be a finite time t', if $t > t'$, then $\sigma(t) \equiv 0$. ■

Proof: Let us prove it by contradiction method. It is supposed that when $t > t'$, $\sigma(t)$ is not always 0.

Assuming that there is a certain moment $\sigma(t) \ne 0$, we might set $\sigma(t) > 0$ by the local scope. From the formula (2), there is $\dot{\mu}(t) = -\gamma$. The integral on both sides about time t is calculated, and we obtained:

$$\begin{aligned} \int_{t'}^{t}\dot{\mu}(\tau)d\tau &= \int_{t'}^{t}(-\gamma)d\tau \\ \mu(t) - \mu(t') &= -\gamma(t - t') \\ \mu(t) &= \mu(t') - \gamma(t - t') \end{aligned} \tag{4}$$

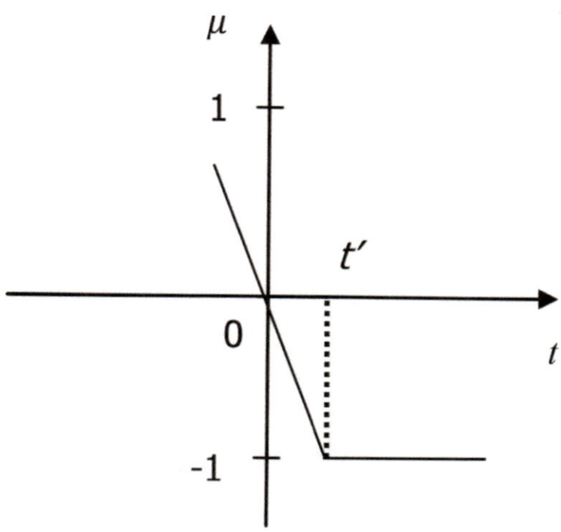

Figure 1.
Local changes of $\mu(t)$ over time.

Knowing from the definition of $\mu(t)$ that $|\mu(t)| \leq 1$, and $\dot{\mu}(t) = -\gamma \leq 0$, so at a certain moment t_1, as shown in the **Figure 1**, once $\mu(t)$ reaches the value $-sign(\sigma(t))$, there is $\mu(t) \equiv -sign(\sigma(t))$ (i.e. $\dot{\mu}(t) = 0$). Otherwise, it contradicts $\dot{\mu}(t) \leq 0$.

Here, t_1 is found in the following method. Let $t = t_1$, we have

$$\mu(t_1) = \mu(t') - \gamma(t_1 - t') = -1 \tag{5}$$

Then there is

$$\begin{aligned} t_1 &= t' + \frac{1+\mu(t')}{\gamma} \\ &\leq t' + \frac{2}{\gamma} \end{aligned} \tag{6}$$

So when $t > t' + \frac{2}{\gamma}$, there is $t > t_1$, there must be $\mu(t) \equiv -sign(\sigma(t))$. By $\sigma(t) = g(t) + k\mu(t)\int_{t_0}^{t} |e(\tau)| d\tau$, then

$$\begin{aligned} \sigma(t) &= g(t) + k\mu(t)\int_{t_0}^{t} |e(\tau)| d\tau \\ &= g(t) - ksign(\sigma(t))\int_{t_0}^{t} |e(\tau)| d\tau \end{aligned} \tag{7}$$

Then,

$$\begin{aligned} \sigma^2(t) &= \sigma(t)\left[g(t) - ksign(\sigma(t))\int_{t_0}^{t} |e(\tau)| d\tau\right] \\ &= \sigma(t)g(t) - k|\sigma(t)|\int_{t_0}^{t} |e(\tau)| d\tau \\ &\leq |\sigma(t)||g(t)| - k|\sigma(t)|\int_{t_0}^{t} |e(\tau)| d\tau \end{aligned} \tag{8}$$

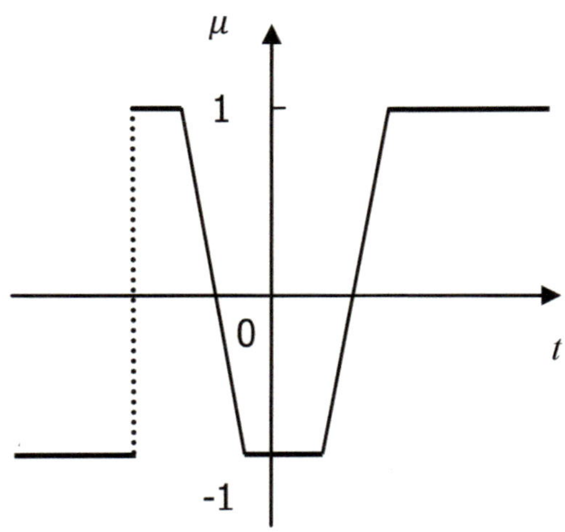

Figure 2.
Overall changes of $\mu(t)$ over time.

From the condition (3), we know that $\sigma(t)^2 < 0$, it leads a contradictory. When $\sigma(t) < 0$, the contradiction can be derived in the same way, and the overall change of $\mu(t)$ will be shown in **Figure 2**. In summary, the conclusion is established. ■

3. Estimation of multiple time-varying parameters based on the new stripping principle

The new stripping principle (NSP) in control theory can be effectively to deal with the interactive influence of nodes in complex network systems. Based on this, we use it to strip the unknown disturbance problem in the extended state observer in the time-varying parameter estimation. Therefore, this section proposes an estimation method for multiple time-varying parameters based on the combination of the NSP with the state observer.

3.1 The statement of the problem

The following system with multiple parameters, and the system itself is highly coupled, as shown in the following system:

$$\begin{cases} \dot{x}_1 = f_1(x_1, x_2, \dots, x_n, \theta_1) \\ \dot{x}_2 = f_2(x_1, x_2, \dots, x_n, \theta_2) \\ \quad \vdots \\ \dot{x}_n = f_n(x_1, x_2, \dots, x_n, \theta_n) \\ y = x = [x_1, x_2, \dots, x_n]^T \end{cases} \tag{9}$$

where $\dot{x}$ means the derivate function of x with respect to time t. For this type of coupling problem, the method of NSP was proposed ([7, 9–11]). If x_i $(i = 1, 2, \cdots, n)$ is regarded as the interconnected nodes in the network, then the system (9) represents each different network node and the relationship among them. This kind of problems are common in real life, such as WeChat, QQ, Sina Weibo etc. in social networking tools. For example, a complex social system is formed through mutual attention and friendship between people, so here the individual x_i is one of them, $f_i(x_1, x_2, \dots, x_n, \theta_i)$ is the interaction (such as relations or research works) among x_i, there are also other network problems like this.

There are always more or less unknowns in the modeling of such problems, which need to be estimated by using known information, which is the multi-parameter estimation problem to be analyzed in this section.

Assuming that $\frac{\partial f_i}{\partial \theta_i} \neq 0$, the following subsection uses the method of combining state observer with NSP to study the parameter estimation method. Specific parameter estimation methods, the convergence analysis and simulation research are described in detail in the following subsections respectively.

3.2 Parameter estimation method

This subsection discusses the parameter estimation method based on the combination of the state observer with the new stripping principle. We need to use the state observer to solve the parameter estimation problem, and we consider the time-varying parameters and time-invariant parameters as well.

Firstly, we extend the unknown parameters in the system (9) to states:

$$\begin{cases} \dot{x}_i = f_1(x_1, x_2, \dots, x_n, x_{n+i}) \\ \dot{x}_{n+i} = g_i(t), (i = 1, \dots, n) \end{cases} \tag{10}$$

That is to say, the parameter θ_i is extended to the system state x_{n+i}, $i = 1, \dots, n$. Then a state observer is built for the extended system (10):

$$\begin{cases} \dot{\hat{x}}_i = f_1(\hat{x}_1, \hat{x}_2, \dots, \hat{x}_n, \hat{x}_{n+i}) + l_i(t) \\ \dot{\hat{x}}_{n+i} = l_{n+i}(t), (i = 1, \dots, n) \end{cases} \tag{11}$$

where $l_i(t)$ $(i = 1, \dots, n)$ is the function to be designed. Let the error $\overline{x} = x - \hat{x}$, then the error system of the established state observer is as follows:

$$\begin{cases} \dot{\overline{x}}_i = f_1(x_1, x_2, \dots, x_n, x_{n+i}) - f_1(\hat{x}_1, \hat{x}_2, \dots, \hat{x}_n, \hat{x}_{n+i}) - l_i(t) \\ \dot{\overline{x}}_{n+i} = g_i(t) - l_{n+i}(t), (i = 1, \dots, n) \end{cases} \tag{12}$$

Next, we need to design $l_i(t)$ to make the error system (12) asymptotically stable. The choice of $l_i(t)$ equivalents to the control problem of the uncertain system (12). We design the control items $l_i(t)(i = 1, \dots, 2n)$ according to the error of the states, so that the error system (12) is asymptotically stable to zero.

Regarding the relevant conclusions of the estimation problem with multiple time-varying parameters in a nonlinear system, we present it in the form of the following theorem and give a stability analysis.

Theorem 1 If the system (10) satisfies f_i (i=1,2,...,n) is differentiable, and $\frac{\partial f_i}{\partial x_{n+i}} \neq 0$ (i=1,2,...,n), take the error feedback $l_i(t)$ in the following form:

$$\begin{cases} l_i(t) = \sum_{j=1}^{n} k_{ij} sign(\overline{x}_j) \\ l_{n+i}(t) = l_{n+i,1}(t) + l_{n+i,2}(t) \\ l_{n+i,1}(t) = k_i \mu_{1i}(t) \int_{t_0}^{t} |\overline{x}_i(\tau)| d\tau \\ l_{n+i,2}(t) = k_{n+i,i} |u_i(t)| \overline{x}_i sign\left(\frac{\partial f_i}{\partial x_{n+i}}\right) \\ i = 1, \dots, n \end{cases} \tag{13}$$

where $u_i(t) = \mu_i(t)(b_{1i}|\epsilon_{i1}(t)|^{\alpha_i} + b_{2i}|\epsilon_{i2}(t)|^{\alpha_i})$, $\mu_{1i}(t)$ is determined by binary control as follows:

$$\dot{\mu}_{1i}(t) = \begin{cases} -\gamma_{1i} sign(\sigma_{1i}(t)), & |\mu_{1i}(t)| \leq 1, \ |\mu_{1i}(t_0)| \leq 1 \\ -\omega_{1i}\mu_{1i}(t), & |\mu_{1i}(t)| > 1 \end{cases} \tag{14}$$

where $\sigma_{1i}(t) = g_i(t) - l_{n+i,1}(t) = \dot{\overline{x}}_{n+i}(t) + l_{n+i,2}(t) \doteq k_{d_i}\ddot{x}_i(t) + l_{n+i,2}(t)$. $\mu_i(t)$ is determined by binary control as follows:

$$\dot{\mu}_i(t) = \begin{cases} -\gamma_i sign(\sigma_i(t)), & |\mu_i(t)| \leq 1, |\mu_i(t_0)| \leq 1 \\ -\omega_i \mu_i(t), & |\mu_i(t)| > 1 \end{cases} \tag{15}$$

where $\sigma_i(t) = \epsilon_{i1}(t) + c_i\epsilon_{i2}(t)$, $k_{(n+i),i}$ is greater than 0, so here it is set that the upper bound of $k_{(n+i),i}|u_i| / \left(\frac{\partial f_i}{\partial x_{n+i}}\right)$ is K_m. k_{d_i}, k_{ij}, ω_{1i}, ω_i, c_i are all greater than 0, γ_i, b_{1i} and b_{2i} are design parameters in the formula (5.16) and formula (5.17) in Ref. [6], $0 < \alpha_i \leq 1$, $i, j = 1, \dots, n$. The design parameters k_i, γ_{1i} respectively satisfy:

$$k_i > \sup_{t \geq t_0} \left| \frac{g_i(t)}{\int_{t_0}^{t} |\overline{x}_i(\tau)| d\tau} \right| \quad i = 1, 2, \dots, n \tag{16}$$

$$\gamma_{1i} > \sup_{t \geq t_0} \left| \frac{k_i |\overline{x}_i(t)| + \dot{g}(t)}{k_i \int_{t_0}^{t} |\overline{x}_i(\tau)| d\tau} \right| \quad i = 1, 2, \dots, n \tag{17}$$

Then the system (11) can be used as an observer of the extended system (10), and there is the results as follows:

$$\lim_{t \to \infty} \hat{x}_i(t) = x_i(t), \quad i = 1, \dots, 2n. \tag{18}$$

■

Now we prove the theorem 1 according to the lemma 1.

If $f_i(x_1, x_2, \dots, x_n, x_{n+i}) - f_i(\hat{x}_1, \hat{x}_2, \dots, \hat{x}_n, \hat{x}_{n+i})$ can be approximately expanded to $\sum_{j=1}^{n} \frac{\partial f_i}{\partial x_j} \overline{x}_j + \frac{\partial f_i}{\partial x_{n+i}} \overline{x}_{n+i}$ by Taylor expansion. Select proper $l_i(t)$ $(i = 1, \dots, n)$ to restrain the main part of $\frac{\partial f_i}{\partial x_j} \overline{x}_j (\, j = 1, \dots, n)$. Then the error system (12) will become the following ones:

$$\begin{cases} \dot{\overline{x}}_i = \sum_{j=1}^{n} \frac{\partial f_i}{\partial x_j} \overline{x}_j + \frac{\partial f_i}{\partial x_{n+i}} \overline{x}_{n+i} - l_i(t) \\ \dot{\overline{x}}_{n+i} = g_i(t) - l_{n+i}(t), \quad (i = 1, \dots, n) \end{cases} \tag{19}$$

At this point, the problem is transformed into a control problem of the system (19).

Known from the conditions that $\sigma_{1i}(t) = g_i(t) - l_{n+i,1}(t) = \dot{\overline{x}}_{n+i} + l_{n+i,2}(t)$ (i=1,2,...,n). Because the actual value of the parameter $\theta_i (i = 1, 2, \dots, n)$ is unknown, so $\dot{\overline{x}}_{n+i}$ $(i = 1, 2, \dots, n)$ are also unknown. In order to estimate unknown parameters, it is necessary to find the equivalent or related quantities of $\dot{\overline{x}}_{n+i}$. From the formula (19), $\overline{x}_{n+i}$ is related to $\dot{\overline{x}}_i$, And $\frac{\partial f_i}{\partial x_{n+i}}$ is bounded, the observer designed here with $k_{d_i} \ddot{\overline{x}}_i$ instead of $\dot{\overline{x}}_{n+i}$in $\sigma_{1i}(t)$, where k_{d_i} is a design parameter. $\dot{\overline{x}}_{n+i}$ can also be replaced with other function forms of $\ddot{\overline{x}}_i$.

According to the lemma 1, when the conditions (16) and (17) are satisfied, it can be obtained that $g_i(t) - l_{n+i,1}(t) \equiv 0$ $(i = 1, \dots, n)$ when the time reaches a certain moment, so the error system (12) is approximately equivalent to the following system without unknown $g(t)$:

$$\begin{cases} \dot{\overline{x}}_i = \frac{\partial f_i}{\partial x_{n+j}} \overline{x}_{n+j} \\ \dot{\overline{x}}_{n+i} = -l_{n+i,2}(t), (i = 1, \dots, n) \end{cases} \tag{20}$$

Since $\sigma_i(t) = \epsilon_{i1}(t) + c_i \epsilon_{i2}(t)$ (i=1,2,...,n), according to the lemma 1, when the design parameters b_{1i}, b_{2i}, γ_i (i=1,2,...,n) satisfied the conditions that the formula (5.16) and formula (5.17) in Ref. [6], and the time is greater than a certain moment, there will be $\sigma_i(t) = \epsilon_{i1}(t) + c_i \epsilon_{i2}(t) = 0$, namely:

$$\overline{x}_i + c_i \dot{\overline{x}}_i = 0 \quad (i = 1, \dots, n) \tag{21}$$

For analyzing the stability of the error systems, the following Lyapunov function for the system (20) were constructed:

$$V_i = \frac{1}{2}\left(K_m \overline{x}_i^2 + \overline{x}_{n+i}^2\right) \ (i = 1, \dots, n) \tag{22}$$

It is easy to know that, except for the origin, $V_i > 0 \ (i = 1, \dots, n)$. Let us analyze the derivative function of V_i with respect to time,

$$\begin{aligned} \dot{V}_i &= K_m \overline{x}_i \dot{\overline{x}}_i + \overline{x}_{n+i} \dot{\overline{x}}_{n+i} \\ &= K_m \overline{x}_i \dot{\overline{x}}_i + k_{n+i,i} |u_i| \overline{x}_i sign\left(\frac{\partial f_i}{\partial x_{n+i}}\right) \overline{x}_{n+i} \\ &= K_m \overline{x}_i \dot{\overline{x}}_i + k_{n+i,i} |u_i| \overline{x}_i \dot{\overline{x}}_i / \left(|\partial f_i / \partial x_{n+i}|\right) \\ &= \left[K_m - k_{n+i,i} |u_i| / \left(|\partial f_i / \partial x_{n+i}|\right)\right] \overline{x}_i \dot{\overline{x}}_i \\ &(i = 1, \dots, n) \end{aligned} \tag{23}$$

From the formula (21),

$$\dot{V}_i = -c_i \left[K_m - k_{n+i,i} |u_i| / \left|\frac{\partial f_i}{\partial x_{n+i}}\right|\right] \dot{\overline{x}}_i^2 \ (i = 1, \dots, n) \tag{24}$$

Known by the condition $k_{n+i,i}|u_i| / \frac{\partial f_i}{\partial x_{n+i}}$ has an upper bound K_m, then $\dot{V}_i < 0 \ (i = 1, \dots, n)$.

In summary, when the time is greater than a certain moment, $\hat{x}_i, \hat{x}_{n+i}$ can be used as the estimation of $x_i, \theta_i \ (i = 1, \dots, n)$ respectively. During this progress, there is nothing to do with the specific form of $g(t)$. ■

Remark 1 When the parameter is a time-invariant parameter, it is easy to prove that the theorem 1 still works. Because at this time the expanded states $g_i(t) = 0 (i = 1, \dots, n)$ in the (10), then we can take $l_{n+i,1}(t) = 0 \ (i = 1, \dots, n)$ in our control law.■

The subsection focuses on the estimation problem of multiple time-varying parameters in general nonlinear systems. A parameter estimation method based on the combination of the state observer with the new stripping principle is given. Stability analysis is also carried out. The following simulation studies further verify the effectiveness of the parameter estimation method proposed in this subsection.

3.3 Simulation analysis

This subsection simulates the parameter estimation method proposed in the previous subsection. We have studied the estimation of a single time-varying and time-invariant parameter, and the estimation of multiple time-varying and time-invariant parameters in a dynamic system respectively. We also consider whether the observation contains observation noise or not. Further verify the robustness of the parameter estimation method.

3.3.1 Single parameter estimation simulation analysis

Example 1 We choose the nonlinear system as follow (that is, example 2 in Ref. [5]):

$$\dot{x} = -|x|\theta + x + cos\theta \tag{25}$$

Here, we assume that the true value of the unknown parameter changes with time $\theta = 1 + sin\ (2t)$, and the initial state of the system is $x(0) = 2$.

According to the system (25), there is $\frac{\partial f}{\partial \theta} = -|x| - sin\theta$. The situation that x and θ are both 0 almost never exists, so it can be considered that the condition $\frac{\partial f}{\partial \theta} \neq 0$ is established.

Next, the extended state system based on the parameter estimation method is described as below:

$$\begin{cases} \dot{x}_1 = -|x_1|x_2 + x_1 + cosx_2 = f(x_1, x_2, t) \\ \dot{x}_2 = g_1(t) \end{cases} \tag{26}$$

We design its observer as follows:

$$\begin{cases} \dot{\hat{x}}_1 = f(\hat{x}_1, \hat{x}_2, t) + k_{11}sign(y - \hat{x}_1) \\ \dot{\hat{x}}_2 = k_1\mu_{11}\int_{t_0}^{t}|\overline{x}_1(\tau)|d\tau + k_{21}|u_1|\overline{x}_1 sign\left(\left.\dfrac{\partial f}{\partial x_2}\right|_{\hat{x}_1,\hat{x}_2}\right) \end{cases} \tag{27}$$

The design of μ_{11} and u_1 is shown in the theorem 1, here we will not repeat them again.

Firstly, we consider the case where the observation does not contain noise. Set the design parameters in the simulation analysis as $k_1 = 0.1$, $k_{d_1} = 0.1$, $k_{11} = 40$, $k_{21} = 10$, $b_{11} = 1$, $b_{12} = 10$, $\alpha_1 = 0.5$, $c_1 = 1$, $\omega_{11} = \omega_1 = 3$, $\gamma_{11} = \gamma_1 = 10$, $\mu_{11}(0) = \mu_1(0) = 0$. Suppose that the initial state of the state observer is $(0, 0)$. We get the estimation of the states and parameters and the estimation errors are shown in the **Figure 3**, the estimation results are satisfied. And after a certain period of time (for example, this simulation is about 7 seconds), the estimated errors of the states and parameters can be controlled within 10^{-2}.

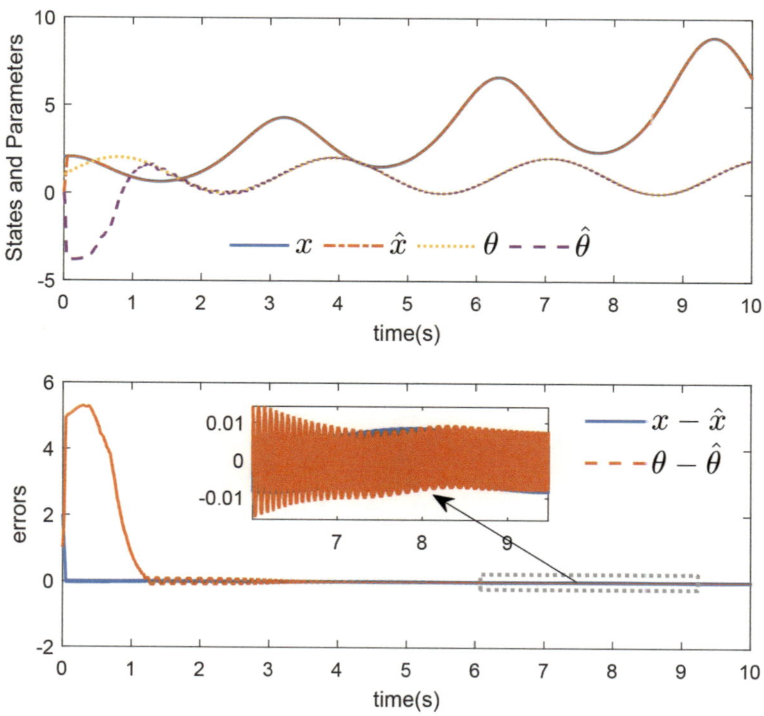

Figure 3.
The case with single parameter and the observation without noise: states, time-varying parameters estimation and estimation errors based on NSP.

Consider when the observation of the system (25) contains noise, for example, there is noise in the observation that obeys uniformly distributed in $[-0.001, 0.001]$, that is, $y(t) = x(t) + \epsilon(t)$, where $\epsilon(t) \sim U[-0.001, 0.001]$. Under these circumstances, design parameters are still taken as $k_1 = 0.001$, $k_{d_1} = 0.01$, $k_{11} = 40$, $k_{21} = 10$, $b_{11} = 1$, $b_{12} = 10$, $\alpha_1 = 0.5$, $c_1 = 1$, $\omega_{11} = \omega_1 = 3$, $\gamma_{11} = \gamma_1 = 10$, $\mu_{11}(0) = \mu_1(0) = 0$, and suppose that the initial state of the state observer is $(0, 0)$. The estimation errors of the state and parameter are shown in the **Figure 4**. Where the estimation error of the state is 10^{-3}, which is larger than the estimation error without noise. The parameter estimation error controlled within 2×10^{-2} is larger than the parameter estimation error without noise as well.

Previously, we studied the estimation problem based on the principle of disturbance stripping for the estimation of a single time-varying parameter, and then we will analyze the situation that the unknown parameter does not change with time.

Example 2 This example is still focusing on the nonlinear system of the system (25):

$$\dot{x} = -|x|\theta + x + cos\theta \tag{28}$$

It is assumed here that the true value of θ is a constant $\theta = 1$ that does not change with time, and the initial state value is $x(0) = 2$.

In the simulation analysis, the parameters are time-invariant, so $g_1(t) = 0$, the feedback item l_{21} can be ignored, and the design parameters $k_1 = 0$, $k_{d_1} = 0$, $k_{11} = 40$, $k_{21} = 10$, $b_{11} = 1$, $b_{12} = 10$, $\alpha_1 = 0.5$, $c_1 = 1$, $\omega_{11} = \omega_1 = 3$, $\gamma_{11} = \gamma_1 = 10$, $\mu_{11}(0) = \mu_1(0) = 0$. Suppose that the initial state of the state observer is $(0, 0)$. The state and parameter estimation and estimation errors are shown in the **Figure 5**.

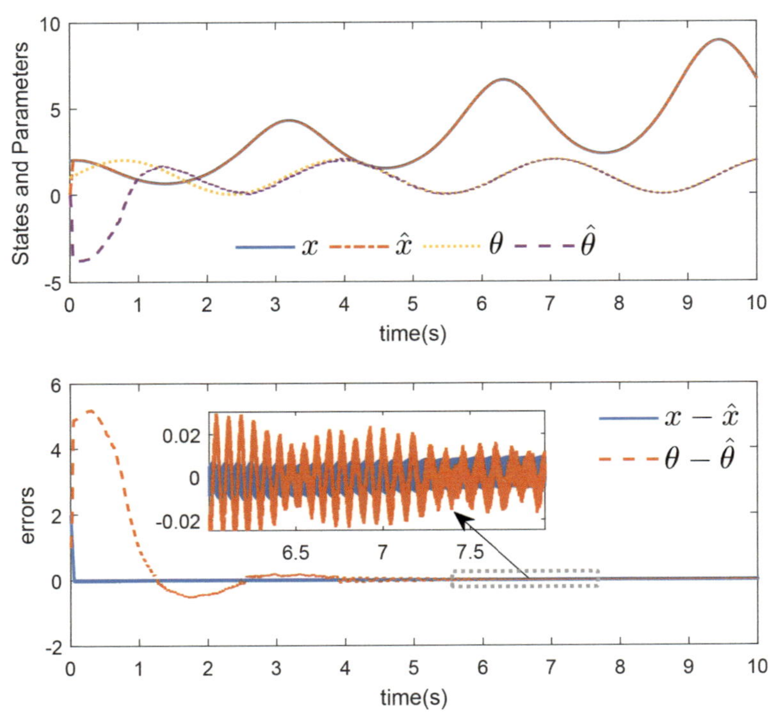

Figure 4.
The case with single parameter and the observation with noise: states, time-varying parameters estimation and estimation errors based on NSP.

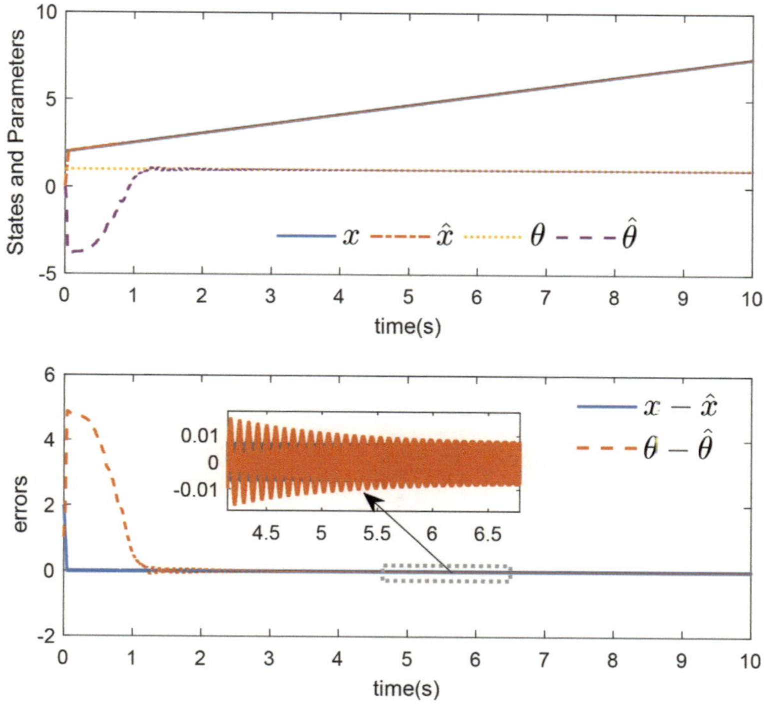

Figure 5.
The case with single parameter and the observation without noise: states, time-varying parameters estimation and estimation errors based on NSP.

The simulation shows that the state and parameters have close to the true value within 1 second, the estimation error can be controlled within 10^{-2} within 5 seconds, and the state estimation converges to the true state value faster due to the effect of error feedback.

When there is noise in the observation of the system (25), for example, the observation contains uniformly distributed noise that obeys $[-0.001, 0.001]$. That is, $y(t) = x(t) + \epsilon(t)$, where $\epsilon(t) \sim U[-0.001, 0.001]$. Under these circumstances, the design parameters in simulation analysis are still taken as $k_1 = 0$, $k_{d_1} = 0$, $k_{11} = 40$, $k_{21} = 10$, $b_{11} = 1$, $b_{12} = 10$, $\alpha_1 = 0.5$, $c_1 = 1$, $\omega_{11} = \omega_1 = 3$, $\gamma_{11} = \gamma_1 = 10$, $\mu_{11}(0) = \mu_1(0) = 0$. Suppose the initial state of the state observer is $(0, 0)$. The estimated errors of the parameters and states are shown in **Figure 6**, the estimation error of the state is 10^{-3}, which is more than the estimation error without noise. However, the parameter estimation error is larger than the parameter estimation without noise, but the estimation error can still be controlled within 2×10^{-2}.

In summary, this subsection studies the application of parameter estimation methods based on the combination of NSP with state observer in the estimation of single parameters of nonlinear systems. This subsection not only analyzed the two cases of time-invariant and time-varying parameters through simulation, but also analyzed the situation that the observations of the system include observation noise. In these simulation studies, based on the preliminary adjusted design parameters, when analyzing the time-varying and time-invariant parameters, and the presence or absence of observation noise, the design parameters were basically not changed, but the simulation results show that the state and parameters in the observer (27) can asymptotically converge to the true value. These studies show the feasibility and robustness of the combination of the state observer with the stripping principle in the single parameter estimation of nonlinear systems.

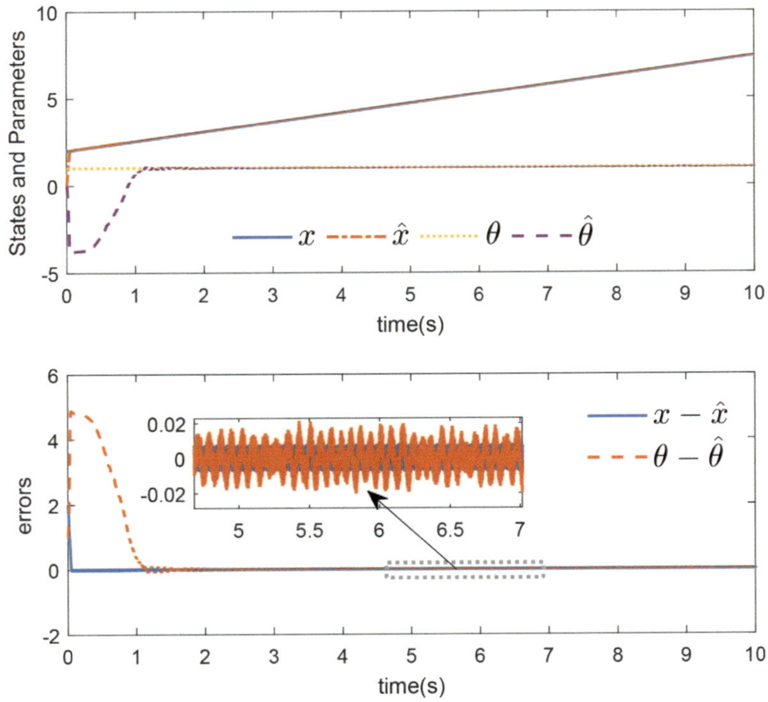

Figure 6.
The observation contains noise: Statse, time-invariant parameters estimation and estimation errors based on NSP.

3.3.2 Multiple parameter estimation simulation analysis

This subsection will study the simulation results with multiple parameter estimates in the dynamic process.

Example 3 Consider the following nonlinear system with two unknown parameters:

$$\begin{cases} \dot{x}_1 = 5\theta_1 x_2 \\ \dot{x}_2 = -\theta_2 cos x_2^2 - 3x_1 \\ y_1 = x_1 \\ y_2 = x_2 \end{cases} \tag{29}$$

Here, we assume that the true value of the unknown parameter changes with time $\theta_1 = sin\,(2t)$, $\theta_2 = cos\,(2t)$, and $\theta_1(0) = 1$, $\theta_2(0) = 1$, the initial state of the system is $x(0) = (5.4, -1.4)$. In this example, θ_1, θ_2 are unknown parameters. According to the parameter estimation method, the unknown parameters are extended to the state, and we obtain the following extended state system:

$$\begin{cases} \dot{x}_1 = 5x_3 x_2 \\ \dot{x}_2 = -x_4 cos x_2^2 - 3x_1 \\ \dot{x}_3 = g_1(t) \\ \dot{x}_4 = g_2(t) \end{cases} \tag{30}$$

where $g_1(t)$,$g_2(t)$ are unknown functions of time t.

The state observer of the above system is established as follows:

$$\begin{cases} \dot{\hat{x}}_1 = 5\hat{x}_3\hat{x}_2 + l_1 \\ \dot{\hat{x}}_2 = -\hat{x}_4 \cos \hat{x}_2^2 - 3\hat{x}_1 + l_2 \\ \dot{\hat{x}}_3 = l_3 = l_{31} + l_{32} \\ \dot{\hat{x}}_4 = l_4 = l_{41} + l_{42} \end{cases} \tag{31}$$

Let $\overline{x}_1 = y_1 - \hat{x}_1, \overline{x}_2 = y_2 - \hat{x}_2$, where l_i is set as follows:

$$\begin{cases} l_1 = k_{11} sign(\overline{x}_1) + k_{12} sign(\overline{x}_2) \\ l_2 = k_{21} sign(\overline{x}_1) + k_{22} sign(\overline{x}_2) \\ l_{31} = k_1 \mu_{11} \int_0^t |\overline{x}_1(\tau)| d\tau \\ l_{32} = k_{31} |u_1| \overline{x}_1 sign\left(\frac{\partial f_1}{\partial x_3}\right) \\ l_{41} = k_2 \mu_{12} \int_0^t |\overline{x}_2(\tau)| d\tau \\ l_{42} = k_{42} |u_2| \overline{x}_2 sign\left(\frac{\partial f_2}{\partial x_4}\right) \end{cases} \tag{32}$$

The design of μ_{11}, μ_{12}, u_1 and u_2 is shown in the theorem 1.

We consider the case that the observation does not contain noise first. By using the design of the aforementioned observer (31), design parameters in simulation analysis are as $k_1 = k_2 = 0.01$, $k_{d_1} = k_{d_2} = 0.1$, $\omega_{11} = \omega_{12} = 3$, $\gamma_{11} = \gamma_{12} = 10$, $k_{11} = 15$, $k_{21} = 0.1$, $k_{12} = 0.1$, $k_{22} = 1$, $k_{31} = 50$, $k_{42} = 50$, $b_{11} = b_{21} = 15$, $b_{12} = b_{22} = 25$, $\alpha_1 = \alpha_2 = 0.5$, $c_1 = c_2 = 5$, $\omega_1 = 5$, $\omega_2 = 0.55$, $\gamma_1 = 10$, $\gamma_2 = 150$, $\mu_1(0) = \mu_2(0) = 0$. Suppose the initial state of the state observer is $(0, 0)$. We obtain the following states, parameters estimation and estimation errors as shown in **Figure 7**.

When the observation of the system (29) contains noise, for example, the observation contains noise that obeys uniformly distribute in $[-0.001, 0.001]$, namely $y(t) = x(t) + \epsilon(t)$, where $\epsilon(t) \sim U[-0.001, 0.001]$. In this case, the design parameters are the same as the above, and the estimated error of the states and parameters are shown in **Figure 8**.

Simulation results in **Figures** 7 and **8** show that the observer designed in this section is applicable to the estimation of time-varying parameters and it has certain robustness to noise. The estimation error of the state is similar either with or without observation noise. For parameter estimation, when there is no noise in the observation, the parameter estimation error is controlled within 5×10^{-2}, but when there is noise in the observation, the parameter estimation effect of θ_2 is not ideal, and the design parameters need to be adjusted appropriately to obtain the more accurate estimation value.

We have studied the estimation of multiple time-varying parameters based on the principle of disturbance stripping above. The following will analyze the situation where the unknown parameters do not change with time.

Example 4 This example is still researching the system (29):

$$\begin{cases} \dot{x}_1 = 5\theta_1 x_2 \\ \dot{x}_2 = -\theta_2 cos x_2^2 - 3x_1 \\ y_1 = x_1 \\ y_2 = x_2 \end{cases} \tag{33}$$

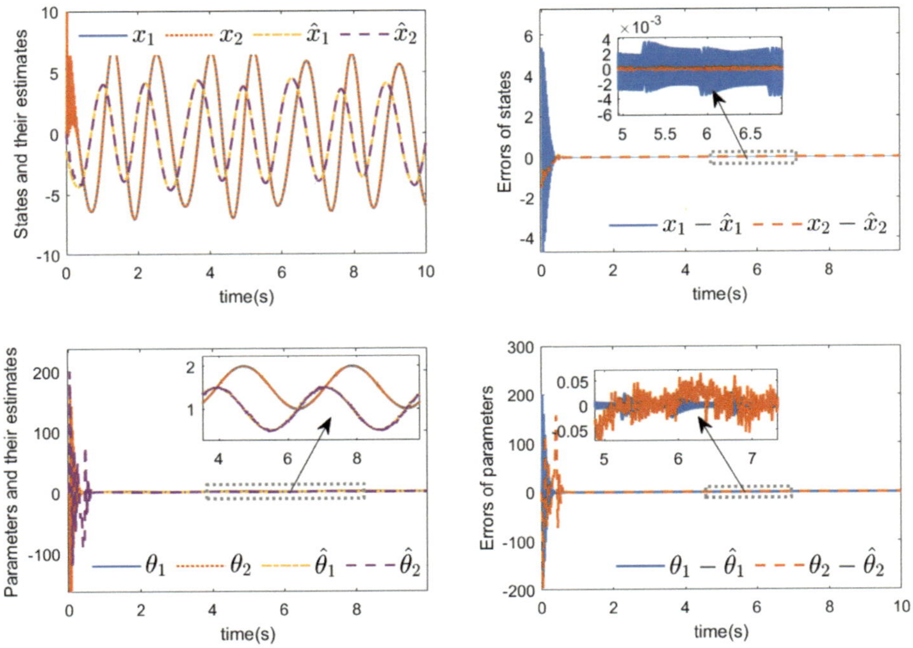

Figure 7.
The case with two parameters and the observation without noise: states, time-varying parameters estimation and estimation errors based on NSP.

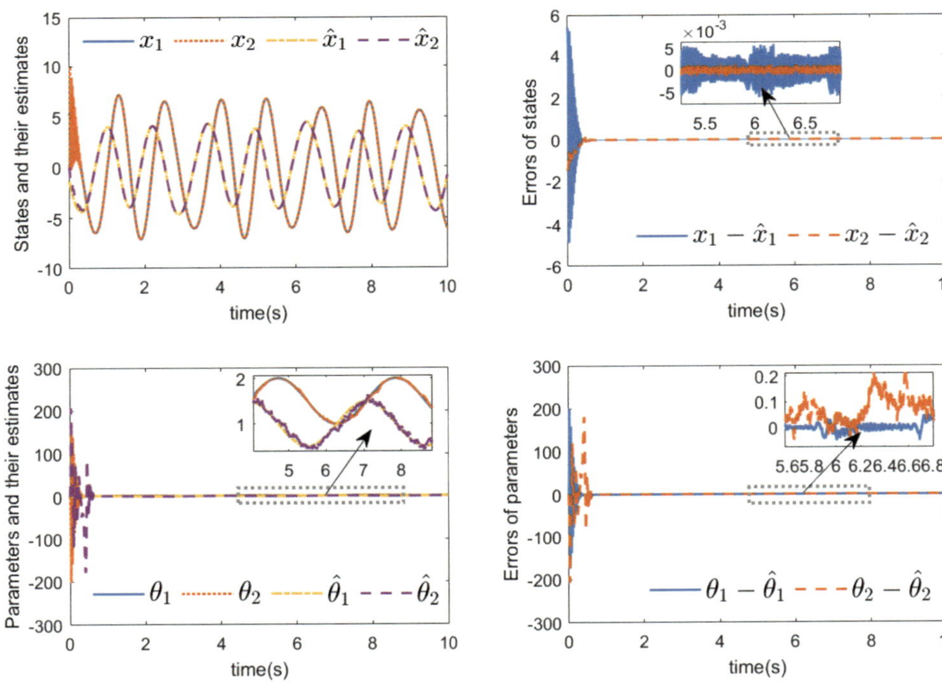

Figure 8.
The case with two parameters and the observation with noise: states, time-varying parameters estimation and estimation errors based on NSP.

Here we assume that the true value of the unknown parameter does not change with time. Suppose that $\theta_1 = 1.4$, $\theta_2 = -1.4$, the initial state of the system is $x(0) = (5.4, -1.4)$.

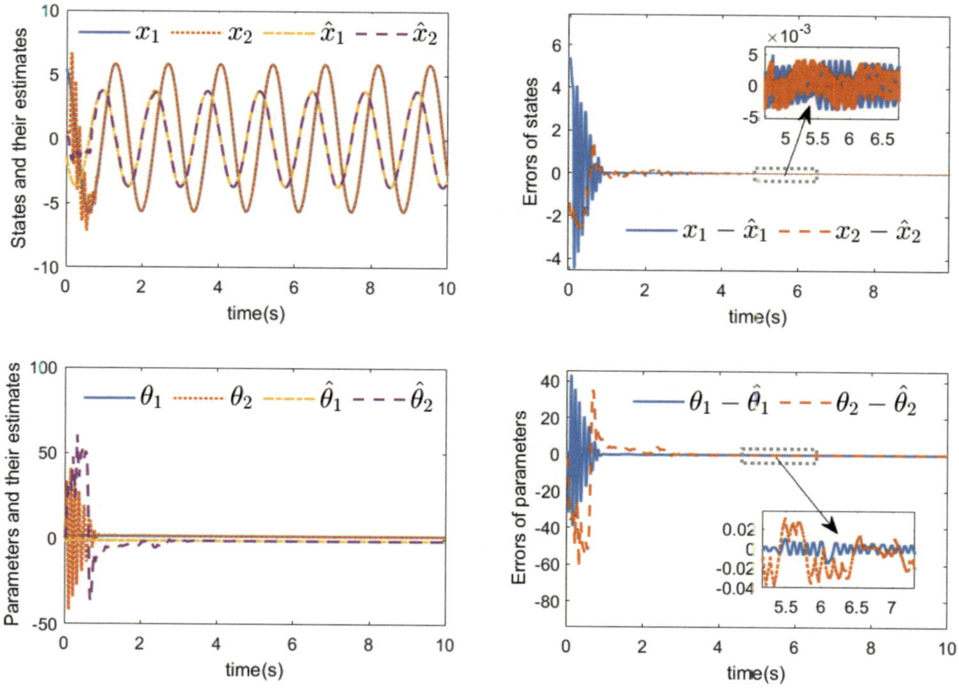

Figure 9.
The case with two parameters and the observation with noise: states, time-varying parameters estimation and estimation errors based on NSP.

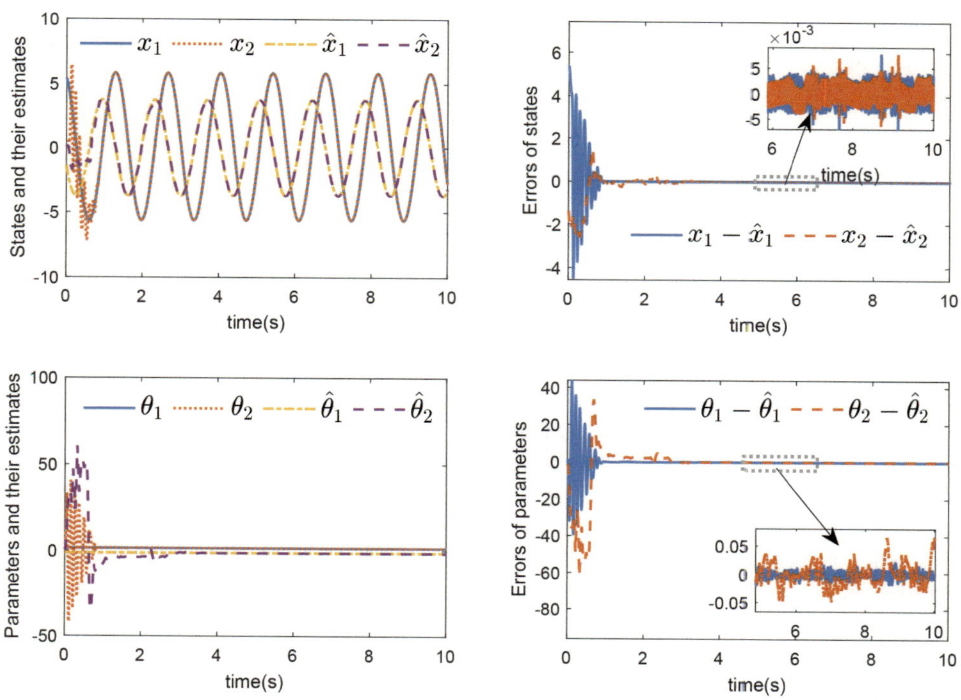

Figure 10.
The observation with noise: The states, time-invariant parameters estimation and estimation errors based on NSP.

For the estimation of time-invariant parameters, $g_1(t) = g_2(t) = 0$, so take $l_{31} = l_{41} = 0$. Take design parameters in simulation analysis $k_1 = k_2 = 0$, $k_{d_1} = k_{d_1} = 0$, $\omega_{11} = \omega_{12} = 0.35$, $\gamma_{11} = \gamma_{12} = 10$, $k_{11} = k_{21} = 10$, $k_{12} = k_{22} = 5$, $k_{31} = k_{32} = 15$, $k_{41} = k_{42} = 10$, $b_{11} = b_{21} = 1$, $b_{12} = b_{22} = 25$, $\alpha_1 = \alpha_2 = 0.1$, $c_1 = c_2 = 6$, $\omega_1 = \omega_2 = 0.35$, $\gamma_1 = 10$, $\gamma_2 = 15$, $\mu_1(0) = \ mu_2(0) = 0$. Suppose the initial state of the state observer is $(0, 0)$. The state and parameter estimation, and estimation error obtained by simulation are shown in **Figure 9**. The simulation result in **Figure 9** shows that the above parameter method is still applicable to the estimation of time-invariant parameters. We can see from the simulation results that the estimation error of the state is controlled within 5×10^{-3}, and the estimation error of the parameter estimation is controlled within 2×10^{-2}, or even better (see **Figure 9** $\theta_1 - \hat{\theta}_1$). If we tune the design parameters properly, we can get a more accurate estimate.

When the observation of the system (29) contains noise, for example, there is noise that obeys uniformly distribute in $[-0.001, 0.001]$, that is, $y(t) = x(t) + \epsilon(t)$, where $\epsilon(t) \sim U[-0.001, 0.001]$. In this case, the design parameters are the same as above, and the estimated errors of the states and parameters are shown in **Figure 10**, where the estimated error of the states are controlled within 5×10^{-3}. The parameter estimation error is larger than the parameter estimation error without noise, but the parameter estimation error can still be controlled within 5×10^{-2}.

In summary, this section analyzes the estimation problem of multiple time-varying parameters in nonlinear systems based on the parameter estimation method combined the observer with the new stripping principle. Simulation research shows that the parameter estimation method proposed this chapter can estimate multiple time-varying parameters (this section only considers the estimation of two parameters), and the time-invariant and time-varying conditions of the parameters in the analysis both illustrate the applicability of the parameter estimation method. In addition, the simulation research on whether there is observation noise in the observations verifies the robustness and feasibility of the parameter estimation method proposed in this section.

4. Conclusions

This chapter studies the state observer method of nonlinear system parameter estimation. When the unknown parameters have explicit expressions, we can use the nonlinear tracking-differentiator-based method to estimate the parameters. The unknown parameters which is relatively non-linear system in nonlinear form or is not easy to express by explicit are main considered in this chapter. According to the different characteristics of the parameters contained in the dynamic process, based on the research of the existing literatures, this chapter proposes a new parameter estimation method based on the state observer and NSP. The parameter estimation method based on the combination of state observer with new stripping principle for dynamic systems containing multiple time-varying parameters. This chapter not only proves the feasibility of the method in theory, but also do the simulations. The simulation results show that the design method can approximate the true value of the parameter within a certain error range. The simulations also consider the presence or absence of observation noise. The simulation results not only show that the parameter estimation method introduced in this chapter is robust to noise, but also show the adaptability of the design parameters. Because it is found in the design parameter adjustment that: adjusting the design parameters within a certain range has little effect on the accuracy of parameter estimation, so in the adjustment of design parameters, according to the characteristics of the error system, the thought and method of control system design can be used to give an approximate value to make the state and the parameter converge, and it can also make fine adjustments to make the estimated error meet the actual demand.

Author details

Jinping Feng[1*†] and Wei Wang[2†]

1 School of Mathematics and Statistics, Henan University, KaiFeng, China

2 School of Mathematics, Renmin University of China, Beijing, China

*Address all correspondence to: fjinping@henu.edu.cn

† These authors contributed equally.

References

[1] Han J Q. The observor of nonlinear system. Journal of Control and Decision. 1990; 5(3):57–60. (In Chinese)

[2] Han J Q. A direct method for dealing with nonlinear control systems. Proceedings of the annual meeting of control theory and application of China Society of automation. 1992: 553–559.(In Chinese)

[3] Wang W, Han J Q. A method of estimating parameters for nonlinear systems. Journal of Control and Decision.1993; 8(3): 161–227.(In Chinese)

[4] Xiang G B. IFTAE optimal control. Beijing: Machinery Industry Press, 1986. (In Chinese)

[5] Wang W, Han J Q. Identification of time-varying parameters in nonlinear systems. Proceedings of Control Theory and its Application Conference. 1993: 300–304.(In Chinese)

[6] Wang W, Han J Q. A method of parameter self-tuning of nonlinear PID controller – binary control system method. Proceedings of Chinese Control and Decision Conference. 1993: 72–77.

[7] Wang W., The stripping principle to the system synthesis for a kind of complex networked systems. *Advanced Materials Research*, 2011, 220: 999–1002.

[8] Han J Q. Nonlinear PID controller. ACTA AUTOMATICA SINICA. 1994; 20(4):487–490. (In Chinese)

[9] Wang W., The Output Tracking Control for a Kind of Uncertain Networked Systems. *Proceedings of the 8th World Congress on Intelligent Control and Automation*, Jinan, China, 2010: 4407–4412.

[10] Wang W., The stability control for a kind of uncertain networked systems by using a new separation principle. *Proc. of the IEEE International Conference on Control and Automation*, Xiamen, China, June 9–11, 2010: 790–795.

[11] Wang W., The synchronization control for a kind of complex dynamical networks. *Proc. of 2011 International Conference on Network Computing and Information Security*, Guilin, China, 2011: 157–161.

[12] Wang W. Configuration method of PID controller parameter correction. Proceedings of Chinese Control and Decision Conference. 1995: 1079–1082, 1110.(In Chinese)

[13] Wang W. Configuration method of output tracking control for a class of nonlinear systems. Proceedings of Chinese Control Conference. 1995: 405–409.(In Chinese)